U0053697

ENRICH
PUBLISHING

BRAND YOURSELF

FOR ADMISSION TO TOP US BOARDING SCHOOLS

5 Key Steps for International Students

Marybeth Hodson
With Case Studies by Jennifer Yu Cheng

Published by

Enrich Publishing Ltd.
A Member of Enrich Culture Group
Unit A, 17/F, 78 Hung To Road
Kwun Tong, Kowloon,
Hong Kong

Copyright © 2021 by Enrich Publishing Ltd.
With the title *Brand Yourself for Admission to Top US Boarding Schools : 5 Key Steps for International Students*
By Marybeth Hodson with Case Studies by Jennifer Yu Cheng

Edited by Barbara Cao
Cover and interior image by Elizabeth Hodson and Getty Images
Book design by Marco Wong

All rights reserved. This book, or parts thereof, may not be reproduced in any form or by any means, electronic or mechanical, including photocopying, recording or any information storage and retrieval system now known or to be invented, without prior written permission from the Publisher.

ISBN 978-988-8599-46-2

Published & Printed in Hong Kong.

Contents

ABOUT THE AUTHORS

Marybeth Hodson

Senior Partner & Director of US Boarding School Admissions Counseling of ARCH Education

Marybeth is a Senior Partner and the Director of Boarding School Consultation for ARCH Education in Hong Kong since 2010. In her role, she helps international students plan for their new and exciting experiences as US boarding school students. Marybeth has extensive knowledge of New England prep schools, developed through her role as a Senior Associate Admissions Director and Coordinator of International Admission at one of Connecticut's boarding schools, her time spent visiting the top US schools, and last but not least, as the parent of four boarding school educated children. She has a keen sense of what academic profiles and attributes are sought by prep schools. Her intimate knowledge of these schools helps her students identify the schools where they will excel.

Marybeth holds a Master's Degree in Psychology and is an Associate Member of the Independent Education Consultants Association (IECA). Marybeth has a passion for women's lacrosse, having played the sport in college, founded a successful local youth program in her local town, and coached for a competitive boarding school program. The host of many webinars for ARCH Education, Marybeth also presented on FindingSchool.net about the importance of athletics in the boarding school process.

Marybeth revels in meeting face-to-face with her ARCH team and students during her extended annual trips to Hong Kong and Asia. Her time in Hong Kong also provides another opportunity to meet with visiting boarding school representatives and host events during their international admission tours.

When not visiting boarding schools or traveling to Hong Kong to work with her ARCH team and students, Marybeth enjoys many sports and activities. An avid Nordic and backcountry skier, Marybeth is a winter enthusiast. As a lifetime tennis player and high school state champion, Marybeth now also enjoys pickleball. Summertime brings her love of the seashore and a newfound passion for golf. Above all else, Marybeth enjoys spending time with her husband, four children, two grand dogs, and as many friends as can fit around the dining table.

Jennifer Yu Cheng

CTF Education Group Deputy Vice Chairwoman and Group President; ARCH Education and ARCH Community Outreach Co-Founder; Jennifer Yu Cheng Girls Impact Foundation Founder

Jennifer Yu Cheng is passionate about education and has been committed to promoting educational development for more than a decade. Jennifer is the Deputy Vice Chairwoman and Group President of CTF Education Group (CTFEG), and is responsible for the group's strategy.

Collaboration — becomes more highly valued, especially when considering boarding school. Students of the 21st Century must have the skills needed to work with others to innovate and implement new solutions. Most people today agree that one of the essential skills for 21st Century kids is **Critical Thinking**. Boarding schools also agree that critical thinking is a crucial skill. For example, boarding schools encourage students to engage deeply in reading, foster thoughtful examinations and understanding of the importance of questioning, and discuss and make connections to the outside world. In doing so, boarding schools hope to teach students to develop critical reading and thinking skills. Critical thinking is essential to decision making.

Critical thinking has long been regarded as an essential skill, but it is not enough. The higher-order skill of **Creativity** also needs to be cultivated. Acording to the IBM 2010 Global CEO Study, a survey of chief executives in 33 industries cited "creativity" as the most crucial success factor. A quick search on LinkedIn reveals that "creative" is the most frequently used word found in individual profiles. Traditional academics are still essential, but with the increasing speed and availability of information, the ability to extrapolate and transform information is vital.

Effective **Communication** skills are also vital for boarding school readiness. Today's students need to know how to communicate with the world around them, not just in writing and speech but also using multi-media and technology. Boarding school students need to talk to other students, and interact with adults and students in the classrooms, in the dorms and on the sports fields. Communication skills are necessary to be a good science partner, a good listener in literature discussions, and an engaged conversationalist at sit-down dinner or to negotiate with your roommate to clean up their part of the dorm room.

Collaboration results when students work together, learn to compromise, and understand that there is a "greater good." Because boarding schools are residential communities, students who know how to collaborate can contribute to a group project by building on the strengths and interests of the group, can build consensus in a student government meeting, as well as motivate their team when they are down in the championship game. The most effective collaborators demonstrate a willingness to compromise and understand that they may have to sacrifice parts of their ideas for optimal results. With an appreciation for the value of 21st Century Skills, students from all over the world flock to US boarding schools where the 4C's are a cornerstone of learning environment.

Understanding the American Education System

While most boarding schools are secondary schools, an understanding of the structure of the American education system can inform families about their

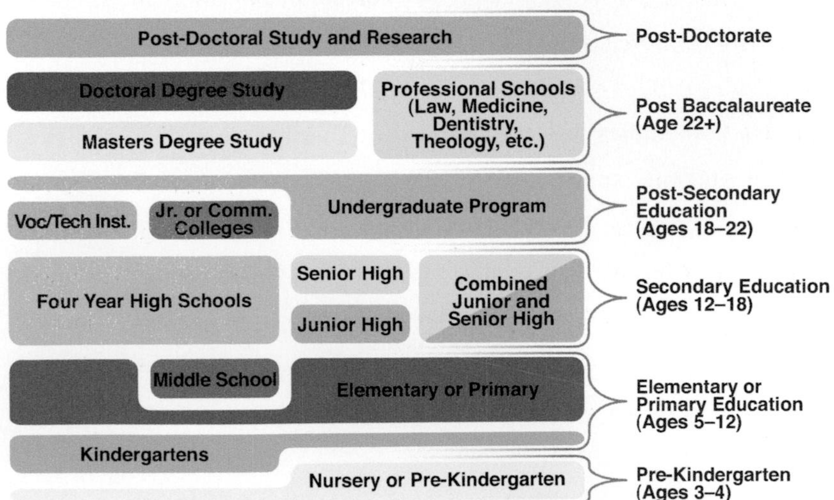

Source: Chart adapted from US Department of Education, National Center for Education Statistics.

options. There are variations, but generally, there are three stages of education in the US, depending on a student's age/grade level: Elementary, Secondary, and Post-Secondary.

The US primary and secondary education systems are generally referred to as K-12, and within this system (Kindergarten to Grade 12) there are public and private schools. Many international families are surprised to learn that the United States does not have a national education system. While the federal government can influence some educational policies based on funding, each state ultimately decides its own educational policies and directives. Additionally, public schools rely heavily on property taxes within their towns/cities to cover most of the schooling costs. Because of this structure, American schools vary considerably based on the educational values and financial wellness of the communities where they are located.

PREFACE

Some American families choose not to attend their local public schools and instead choose to participate in private schools. Private schools do not rely on government funding; they are supported by tuition fees, grants, donations and, in the case of parochial schools, by religious organizations.

Private secondary education to include high school and middle school (ages 12 to 18) will be the primary focus of this book with a more specific target at residential living and learning communities known as boarding schools. Boarding schools have been around for a long time. The United Kingdom has a long history of students attending boarding school dating back to medieval times. Compared to UK boarding schools, US boarding schools would be considered young on a historical timeline with West Nottingham Academy in Maryland being one of the oldest US boarding schools founded in 1744, Phillips Academy Andover established in 1778 and Philips Academy Exeter in 1781, to name a few.

Why Attend a Boarding School?

Families choose a boarding school for a variety of reasons. For some, attending boarding school may be part of their family tradition with parents and grandparents before them having attended. Others are looking for broader opportunities and different values than their local education systems provide. Still others may be looking for greater resources to develop students in areas such as academics, athletics, music, or to help gain life skills and independence.

There are nearly as many types of boarding schools as there are reasons why families decide to attend them. There are traditional co-educational boarding schools, single-sex boarding schools, junior boarding schools, therapeutic boarding schools, 5-day boarding, boarding schools that offer learning support and others that have signature programs. Often, the "reason" to attend is aligned with the type of boarding school a family is targeting.

At boarding schools, students live in dorms on campus and attend classes. Boarding schools are academic, challenging environments where students get support from involved teachers and engaged peers. Because learning occurs in a residential setting, students who attend boarding school are learning 24/7. According to The Association of Boarding Schools (TABS), compared to their private day and public school peers, students at boarding schools spend more time each week doing homework (17 hours vs 9 hours and 8 hours respectively), and 50% of them go on to earn post-graduate degrees, compared to 36% of their private day peers. Boarding schools offer access to a variety of sports, extra-curricular activities, and clubs without the hassle of commuting.

How to Start the Boarding School Journey?

Most families start the boarding school journey with a pre-determined school list based on national rankings with very little understanding about specific schools and the competitive landscape of US boarding schools. All too often, families find that their journey leaves them confused about school fit and what US boarding schools are looking for in candidates for admission.

While the steps covered in this book are referenced with a boarding school target in mind, the key principles apply to admission to any competitive US secondary school that has the ability to enroll international students. Many private and parochial day schools are able to enroll international students through the Student Exchange Visitor Program (SEVP). Additionally, some public high schools, if they have eligibility through SEVP, are open to non-

US citizens to attend for 12 months provided the student pays the cost of the school fees in that district.

Like any meaningful journey, early planning is the key to success. It takes time for students to develop academic readiness, extra-curricular profiles, and character skills to increase their likelihood of being accepted, and ultimately being successful once admitted to boarding school. The planning process must be approached with the fundamental goal being student success, rather than just getting accepted. For a successful process, these two goals go hand in hand. Prepared students not only get admitted, they also ultimately do well throughout boarding school and, consequently, in the university application process and beyond.

An entrepreneurial framework is used throughout this book to help families better visualize the boarding school admission process. Using marketing concepts, such as authentic messaging, brand definition, and customer loyalty, gives context to the priorities that are necessary for a successful boarding school application. This book, ***Brand Yourself for Admission to Top US Boarding Schools : 5 Key Steps for International Students***, is intended to impart a clear understanding of what US boarding schools are looking for in an admissions candidate and how to best prepare for what most families find is a complicated, confusing and time-consuming process.

This guide is based on many years of experience in the boarding school community. My experience began as a parent of 4 boarding school educated children. I then worked as an international admission officer and lacrosse coach at a US boarding school. A decade ago, I transitioned to work as a professional educational consultant, successfully placing students to the most prestigious American boarding schools. While I have worked with students from many countries, including the US, as a Senior Partner and Director of Boarding School Admissions at ARCH Education in Hong Kong, the bulk of my experiences have been placing students from Hong Kong, Mainland China, and other countries in East Asia. However, the principles discussed in this book and the guidance provided can benefit any student wanting to apply boarding school, regardless of which country they are from. By following the 5 Key Steps in this book, families will gain a clear understanding of what types of students boarding schools are looking for, and how to plan to be one of them!

In the final section of the Appendix, ***MY BOARDING SCHOOL PLAN WORKBOOK*** is provided as a download for students to use the tools discussed in this book to develop their own plan to a successful boarding school application.

Step 1 START EARLY

Step 2 PLAN WELL

Step 3 EXECUTE

Step 4 FOLLOW-UP

Step 5 TRANSITION

INTRODUCTION

There are nearly 300 boarding schools in the United States. While many boarding schools are concentrated geographically along the East Coast, with the highest density of schools located in Connecticut and Massachusetts, US boarding schools are represented in 41 of the 50 states in the US, including Hawaii.

Location of US Boarding Schools

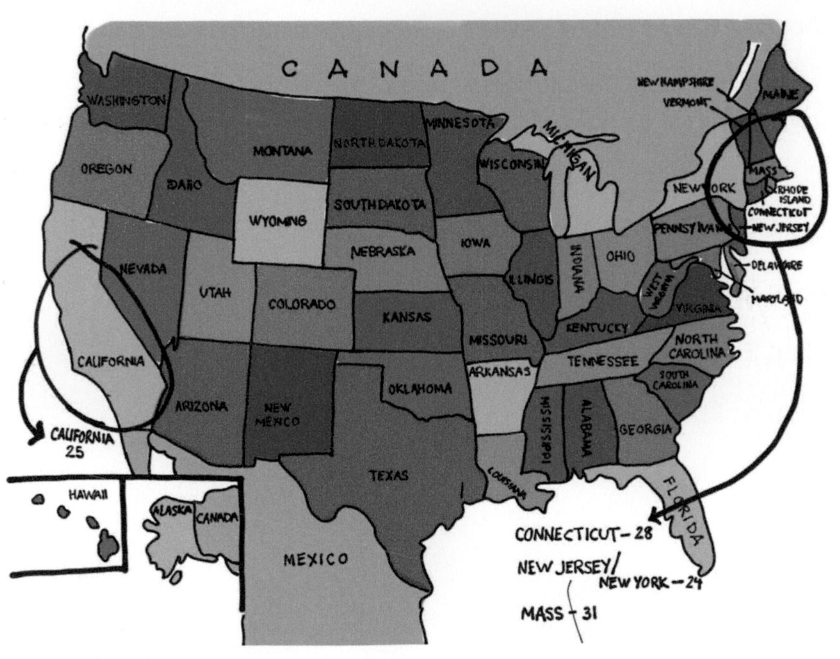

Not all boarding schools are alike. Many factors are to be considered when identifying potential boarding schools, such as selectivity, size, location, special programs, and activities. While boarding schools are known to have excellent academics, competitive athletics, and a tight-knit community, many boarding schools further distinguish themselves with signature programs and unique facilities.

Are You a Fit for a Particular Boarding School?

It seems like common sense that all boarding schools are looking for students who are prepared for the academic challenges required to succeed in these highly selective schools. Yet, alongside academic readiness, schools are also looking for students who can have an immediate impact or those that demonstrate a propensity to be part of and contribute to the school community. Schools are looking for students who are resilient to overcome inevitable challenges, cooperative in their interactions and who demonstrate intellectual curiosity and open-mindedness. Fit is essential. The intent of this book is to focus on the preparation necessary to gain admission to boarding school, as well as the skills necessary to succeed once the student arrives.

Boarding School is not for everyone.

Many Asian parents read published ranking lists and acceptances rates on the Internet and naturally apply these statistics to their son or daughter's chances of being admitted to a US boarding school. Published acceptances rates include all students, that is domestic, international, boarding, and day students. Despite the published acceptance rates, as applications continue to rise from the Asian student cohort, it is my experience that actual acceptance rates tend to run well below those that are published, regardless of school selectivity. It is merely a matter of Economics 101: Supply and Demand. There is a greater supply of qualified applicants than ever before, particularly from Mainland China, and yet, there are only so many schools from which to choose.

INTRODUCTION

Many areas of boarding school life should be considered when thinking about whether a student will just manage to get through the experience, or if they will thrive at a US boarding school. Before going through the demanding and expensive task of applying to a US boarding school, a student must first ask themselves if this learning environment is a good fit for them and if they are ready.

STEP 01 START EARLY

Early preparation is essential for students and their families to gain a full understanding of the boarding school process. By starting preparation early, families can grasp the competitive landscape of boarding school admissions and learn about the advantages and disadvantages of different schools for a particular student. Early preparation allows for an understanding of a student's relative strengths and weaknesses and the time to plan to work on the weaknesses and maximize the strengths.

Many families start thinking about applying to boarding school during the spring of a student's 7th-grade year to apply for Grade 9 entry. On the surface, two years seems like ample time for planning. However, application preparation, interviews, and school visits usually occur during the fall of the year a student applies, which in this example is Grade 8. The time frame to remedy or further develop this student's **brand** *and preparation is only a few months.*

Most students benefit from two full years of planning and preparation before the application year. *For students wishing to enter boarding school in Grade 8, ideally, planning begins in the late winter to early spring of Grade 6.*

1.1 READINESS INDICATORS

Readiness should be the first consideration when thinking about boarding school. Students who are ready to attend boarding school have the qualifications and skill sets necessary for acceptance, as well as for success once admitted. Students who are not prepared risk failing, and in the worse cases, end up returning home. Readiness needs to be demonstrated in 3 different areas for a student to truly be prepared for the challenges of attending boarding school. The readiness indicators are **Academic Readiness**, **Social and Emotional Readiness**, and **Extra-Curricular Readiness**.

ACADEMIC READINESS

There are several areas to consider when planning for student success in being admitted to boarding school and beyond. First and foremost, boarding schools are college preparatory institutions, making **Academic Readiness** a starting point for all students who are considering applying to boarding school. Academic Readiness is a broad term and not only covers a student's ability to succeed in classes like literature and calculus but also addresses a student's ability to adjust to the pedagogy of boarding schools' learning environment. Students who thrive at boarding school tend to be students who are intellectually curious and open-minded. US boarding schools are built on the Socratic Method, where students engage in a cooperative form of dialogue and debate. Successful boarding school students are eager to engage in active dialogue in class and consider other points of view. The Harkness Method of discussion-based learning, a hallmark of most American boarding schools, is inherently suited to students who have excellent communication skills, the desire and ability to express their opinions, and have learned to listen to the views of their peers. A willingness to develop critical thinking skills and to conceptualize, analyze, and synthesize information is also crucial to a successful boarding school academic experience. This style of learning can be quite different than many student's current educational environments. Students who have been educated in a more passive nature where the sharing of opinions and asking questions is not valued and often discouraged will likely need support transitioning to a US boarding school. Outgoing, optimistic, and open-minded students often fare better at boarding school than those who are not.

There are several areas to consider as they relate to a student's academic readiness to attend boarding school.

English Proficiency

Arguably the most crucial factor in determining student success in academic readiness starts with **English Proficiency**. Proficiency in speaking, understanding, and writing in English are essential in most boarding schools. Lack of English skills can not only deny a student the opportunity to learn in the classroom, but it can also impact their transition to residential living, engagement with English speaking students and overall adjustment to boarding school life. Students who are not comfortable with English can find themselves having great difficulty in the classroom, spending countless hours on their homework, and isolated from their peers with whom they can't engage.

Most applicants to US boarding schools should have had some exposure to the English language and many students have skills similar to native English speakers. There are many different learning environments from which students apply that can impact English proficiency and, as a result, affect the selectivity of their boarding school choices. Some students attend international schools where all of their classes (except maybe Chinese/Foreign Language) are conducted in English. Other students may attend schools where their classes for part of the day are taught in English, or specific subjects are taught in English. Still other students may attend local schools where there is less availability for instruction in English. Students also supplement their English learning outside of school.

Some boarding schools offer English as a Second Language or a specific English Learner Curriculum for students who English is not their first language and/or English proficiency is not sufficient. For some students, identifying schools with language support programs is a priority when identifying schools with the right fit.

Most boarding school admission officers will be familiar with some of the well-known schools in most international cities. Admission officers are likely to have worked with and admitted students from these specific schools, and thus are quite familiar with the curriculum and standards of English expectations of students at these schools. Familiarity with a particular school allows the admission officers to better predict how well students from the same school will transition to their school. Students who are applying from lesser-known

STANDARDIZED TESTING

Standardized testing requirements vary from school to school, but it is safe to say that most boarding schools will require **Standardized Testing** as a component of their admission process. Typically, the tests that most boarding schools need for international students are the Secondary School Admission Test (SSAT) and/or the Test of English as a Foreign Language (TOEFL). Some schools will accept the Independent School Entrance Examination (ISEE).

Some schools will place more emphasis on admission testing than others, and what one school considers an admissible score may not be acceptable at another. For all schools, standardized test scores are never the most important factor in deciding which students get admitted to a boarding school. When making admission decisions, boarding schools look at many factors that will be discussed in detail throughout this book. However, because admission testing is part of most boarding school applications, it is best to start early to learn about the different tests.

SSAT

Almost all US boarding schools require the Secondary School Aptitude Test, or SSAT. The SSAT contains four different sections: **Vocabulary, Quantitative, Critical Reading**, and a response to a timed **Writing Sample**. Most students from Asia typically do very well on the

Quantitative section of the SSAT, with many earning a full score. While high scores on the Quantitative portion are commendable, the areas of Vocabulary, Critical Reading, and the Writing Sample are usually most challenging and more important to the US Boarding schools. Looking deeper at the language sections, because the Vocabulary section focuses on vocabulary and relationships with words, it can be easily practiced resulting in drastic improvement. Proficiency in the Critical Reading section becomes a more reliable predictor of academic readiness. Not only is Critical Reading designed to test a student's ability to read and understand written English, the reading section also tests whether students read critically, which means examining and interpreting the meaning of a passage. This section is more difficult to practice or train up and, as such, is considered a more reliable test for predicting a student's abilities to function in these areas at boarding school.

The Writing Sample is also strongly considered as a reliable measure of a student's raw writing ability. In this section, students are allowed 25 minutes to answer 1 of 2 writing prompts. The SSAT Board does not score this section: rather, the Writing Sample is sent to schools as an example of a student's writing. Schools not only use the Writing Sample to evaluate a student's ability to write, but schools also look at the Writing Sample as a benchmark to compare application essays.

It takes time to properly prepare for the SSAT, particularly to do well in the Critical Reading and Writing Sample sections, so early preparation is key. Many organizations can help guide students to best prepare for the SSAT. Students are recommeneded to find a test prep organization, if for nothing else than to gain an understanding of how the test works and how to develop a test-taking strategy that best suits them.

As stated previously, a student's grades are more important to boarding schools than high SSAT scores. That being said, since schools do publish the average SSAT's of students who are accepted, meeting specific standards is necessary. High SSAT scores will not be the only factor that gets a student admitted to boarding school.

TOEFL

Because there are varying degrees of exposure to English for international students, schools must have some mechanism in place to try to evaluate readiness to engage in an academic environment where all classes, except second languages, are taught in English. There are several standardized tests to assess language ability. The most widely used by boarding schools is the Test of English as a Foreign Language, or TOEFL. Some schools also accept IELTS (International English Language Testing System). Most schools require a TOEFL from students who have been studying in a school for the past two years where the primary mode of instruction is not English. Some schools will publish their minimum TOEFL requirements on the admission pages of its website. Sometimes schools use this minimum requirement as a pre-screening mechanism, requiring students to submit proof of the test result before being invited to campus for a tour and interview.

Some students who meet minimum TOEFL requirements may still find themselves unable to understand instruction and discourse when in the classroom. Basic comprehension and grammar may not be sufficient to apply to a US boarding school successfully.

When considering English proficiency, students may want to ask themselves:

- ***Am I able to read and comprehend most English texts?***
- ***Does it take me a reasonable amount of time to read and comprehend in English?***
- ***Do I have a high level of competence in writing and speaking, and understand the nuances of English grammar rules?***
- ***Am I comfortable speaking up in class and expressing my opinion in English or offering input to class discussions?***
- ***Do I have a strong command of English vocabulary and functional language?***
- ***Can I follow and understand lectures in English?***

Students who are not able to answer affirmatively to these questions may want to consider avenues to improve their English proficiency as well as target the many excellent boarding schools with English as a Second Language Programs.

English as a Second Language (ESL) or English Language Learners (ELL) types of programs offer students language instruction and often offer curriculum more geared towards a student whose primary language of instruction has not been English. Keep in mind that most students spend a year or two in classes with English support and eventually graduate out to mainstream instruction. Many parents resist considering schools with English support programs thinking that their child's English will automatically improve at boarding school. Parents of students with a weakness in English may insist that their child attend a school that does not have ESL/ELL. However, time and time again, without the necessary foundation in English, students almost unanimously struggle from the beginning. Without remedial instruction, they are unable to catch up to a level of proficiency. Lack of English proficiency acts like a snowball running downhill that builds and builds. The curriculum becomes more demanding as the students advance grade levels. Despite considerable effort, a student who struggles to earn a C in Freshman English, by Grade 11, when the requirements become more difficult, may be failing severely and jeopardizing their college placement outcome. Students who find themselves above their English ability level in class find it very difficult to ever catch up. A student who is placed at the appropriate level of English ability will indeed be able to improve their English and thrive at boarding school.

At the most competitive boarding schools, most students who are competing in the application process will have excellent grades AND good standardized test scores, AND a high level of English proficiency. With so many students who can demonstrate academic readiness, the factors that influence admission decisions go beyond grades and SSAT scores.

SOCIAL AND EMOTIONAL READINESS

Even the most academically prepared and accomplished students can find the transition to boarding school challenging. Everything is new! New school, a new way of learning, new friends, a new country, a new culture, and a new way of living. One of the key factors that students should consider when applying to boarding school is **Social and Emotional Readiness**. They should ask themselves questions like, "Have I had experiences where I have had to overcome challenges? How did I overcome these challenges, and are these skills developed well enough to help me adapt to the challenges that I will certainly face at boarding school?" "Am I able to manage my time and make mature decisions?"

Learning and living at boarding school is a social process, where students do not learn alone, but rather in collaboration with their teachers, peers, and advisors. They live in a boarding school community that operates much like a family, with dorm faculty that serve as in loco parentis, older students that may take on the role of an older sibling, and younger students who can be like little brothers or sisters.

Living on their own, without direct parental support and guidance can be difficult for some students. Independence is an essential attribute for readiness. Like many situations in life, previous experiences can better prepare for future success. The importance of prior experience holds for adjusting to the challenges of living independently. Students who have demonstrated the ability or have previous experience being self-reliant, such as in attending

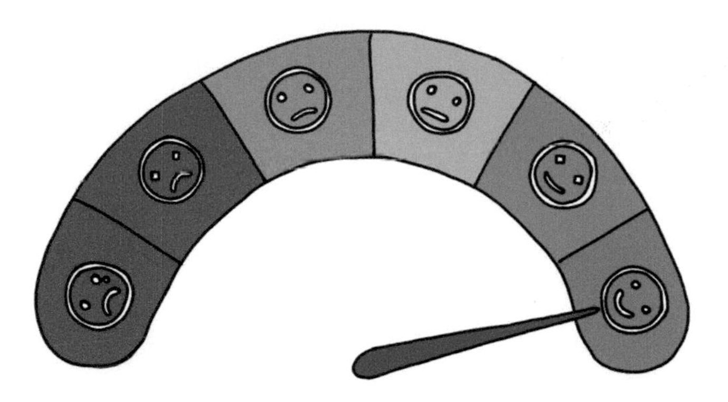

a summer program or exchange program, tend to adjust more readily to the challenges of living independently, thousands of miles away from home in a foreign country. It's important to note here that while boarding schools can help to empower a student's sense of independence, boarding schools are looking for students who are already developing their skills in this area.

A willingness to embrace cultural differences is also a key factor for students who are considering boarding school and demonstrates social and emotional readiness. While the majority of students at most boarding schools are American, with increasing globalization, boarding schools strive to have their schools mirror our globally diverse world. Willingness to not only adjust to American culture and foods but also to accept many different cultures and fully embrace an American boarding school experience is necessary for applicants to be successful.

Because social and emotional readiness is key to success, students and parents need to make sure that the student is socially and emotionally ready to attend boarding school. Determining if a student demonstrates social and emotional readiness for boarding school is not as easy as reading an academic transcript or a standardized test report. Boarding schools try to gain information to assess a student's social and emotional readiness throughout the application process, through teacher recommendations, application essays, and interviews. Admission officers will look at past experiences of living independently, like summer programs or school trips. They look for indicators of how students overcome obstacles and show grit and resiliency. Sometimes it is hard to find the social and emotional readiness indicators, so students should make sure that they think about how they are going to help the boarding schools recognize their social and emotional readiness in the application process.

EXTRA-CURRICULAR READINESS

US Boarding schools have extensive resources, programs, and facilities. They can provide a breadth of programming that is beyond those of most other schools, whether they are public or private, local or international. For most students, regardless of where they are from, boarding school presents new challenges and opportunities. Students who are willing to embrace new experiences like trying out for a new sport, speaking up in class, joining a new club or sitting down at a lunch table with unfamiliar students, are the types of students that boarding schools are seeking.

Extra-Curricular Readiness relates to how well prepared the student is to contribute to a boarding school outside of the classroom. Extra-curricular readiness can be demonstrated in as many ways as there are activities available at boarding schools

All three readiness indicators (Academic, Extra-Curricular and Emotional) need to be present for a student first to get accepted to boarding school and most critically, to succeed at boarding school.

If you have honestly reflected on who you are as a person and feel that indeed you have some of the qualities, attributes, and skills mentioned above, then you are ready to begin the process of learning more about applying to boarding schools in the United States. I invite you to read on.

1.2 BUILD YOUR BRAND

Once it has been determined that the student meets a school's academics readiness indicators, and has the skills and experience base that are important for social and emotional readiness, a boarding school will move on to look for ways to evaluate extra-curricular readiness. The most important questions students should be asking themselves as they are thinking as it relates to Extra-Curricular Readiness in applying to boarding schools are:

- *What is special about me?*
- *How am I going to be remembered?*
- *What impact can I have at a school?*

Successful branding is the key to outdoing competitors and in creating a successful business. For a businessperson, branding is about deciding what is unique about their product, their company, or how the product they are selling is better than other similar products. From there, businesses develop a slogan or brand image that represents the company and best promotes their product(s) in the marketplace. The notion of branding can also be used as a basis for a successful boarding school application. For someone applying to boarding school, your brand is not the latest electronic device you can't live without or the newest brand of sneaker; your brand is **You**!

In developing their brand, students should work to identify and cultivate what are referred to as "X-Factors." X-Factors are distinct activities or interests that set students apart. X-Factors can be anything valued by boarding schools such as athletics, STEM (Science, Technology, Engineering, and Mathematics) or humanities activities, writing, visual arts, drama, community service, music, and more. X-Factors are essential because schools are looking for students who can make an impact on their campus and the world. Students who are engaged in the classroom, activities and other interests will likely encourage others to participate. Involved students build a sense of collaboration and community. Students who have created a good brand by developing their X-Factors are not only more likely to get admitted and do well in the boarding school process, but they also fare much better in the college application process than those who do not.

The next sections will take students through a basic branding plan with the focus on **You** as the brand. Branding is a 2-part process, Part 1: Creating an Authentic Message, Part 2: Promoting Your Brand. There are multiple steps within each part, and these steps will be applied to the boarding school context. **START EARLY** remains the theme for Step 1.

BRANDING PART 1: CREATING AN AUTHENTIC MESSAGE

The first part of building a brand is **Creating an Authentic Message**. To do this, students must pinpoint their mission. For most students, their mission is to demonstrate to a boarding school that he/she is a qualified applicant who can be successful in the classroom and make an impact in their community. For branding to be authentic, it needs to be truthful. To attract a boarding school, students need to project their attributes, talents, and passions, as well as an attitude of resiliency and optimism to experience new opportunities.

Pinpoint Your Mission

In formulating their mission, a businessperson may ask themselves why they want to start a particular business and what goals do they want to achieve. They may ask themselves: how does my business idea differ from similar ideas or products in the marketplace? Likewise, these are questions that students should be asking themselves when thinking about You as your brand:

- ***Why do I want to go to a US boarding school?***
- ***What goals do I hope to achieve at a US boarding school?***
- ***What are my brand's X-Factors?***

Decide How You Want to Be Seen

Americans tend to think big, especially when it comes to shopping. Start thinking of a boarding school as a superstore. A superstore is an extensive retail establishment that offers a variety of products for sale. Stores like Walmart and Target would be considered superstores — they have just about everything you need. A superstore needs many useful products and brands on the shelves for customers to want to come to their store. Like a superstore, boarding schools are looking for different "products" to put on their shelves. They don't want to stock the shelves with the same products, for example, only offering one brand of laundry detergent. They want to fill their shelves with students who have different qualities, or X-Factors. Throughout this book, the analogy of the boarding school as a superstore will be used to help convey the message that boarding schools are looking for students who differentiate themselves, ensuring their brand or "product" is attractive for the boarding school superstore.

Think about your brand image. What are the qualities that differentiate You? Are you a talented math student or a musician? Or maybe you are not just a musician, but you play a unique instrument like the oboe or double bass. Or, perhaps you are a stellar athlete or the captain of your debate team. Boarding schools want to have talented students in various disciplines and at different levels of achievement within those disciplines. They are looking for talented athletes to represent them at the varsity level, as well as eager athletes to fill up some of the less competitive teams. They need students who can represent their school in robotics competitions and participate in the school orchestra. Each of these students is representing their brand: as a musician, an athlete, a builder of robots. Some students may be able to render several brands at the same time, such as a student who is not only talented in academics but is also the star player on the basketball team and leader of a community service club at their current school.

Research and Development of X-Factors

Identifying and developing X-Factors is the part of the boarding school admissions process that takes the most time. Schools are looking for students with diverse interests, skills, backgrounds, and experiences. Think of this as "product development" and consider doing research and development on your "product," **You!** Product development is where following the key principle of **START EARLY** brings the highest return. It takes a while for you to identify and develop your brand or, more specifically, identify and build your X-Factor(s). Conduct research on some of the most common X-Factors that are represented at boarding schools. Honestly evaluate your interests, skills, backgrounds, and experiences in those areas. Identify a strategy to continue to develop the X-Factors that you are passionate about and identify a plan to improve in areas that are not yet a strength.

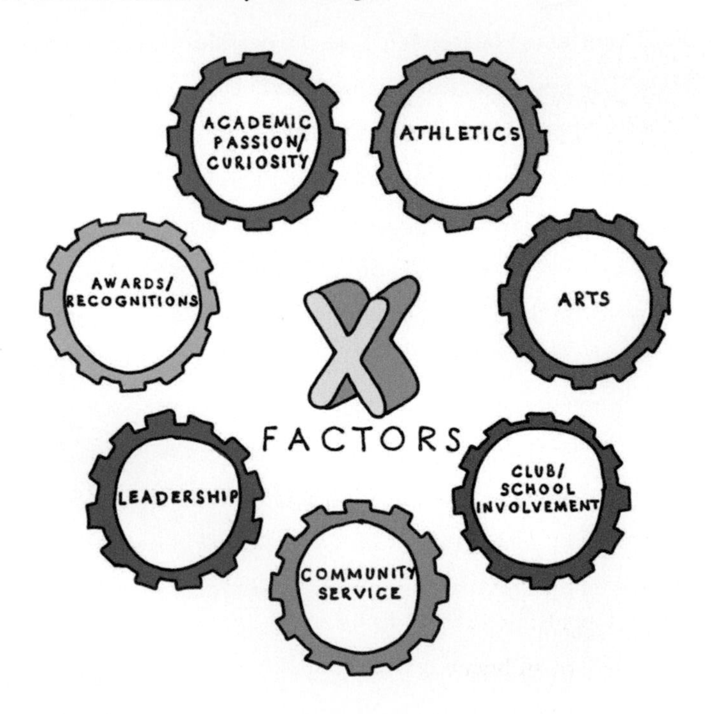

The following list of X-Factors is not all-encompassing but represents those that are commonly found on the shelves or needed at the boarding school superstore.

X-Factor 1: Academic Passion and Curiosity

Boarding schools appreciate candidates that have academic curiosity and passion. They want students who engage in exploring beyond what is required and can ignite group discussions so that everyone benefits and learns.

All schools require students to generally take the same types of classes: English, Math, History, Sciences and usually a Foreign Language. There are many choices within these broad categories, as well as electives. If a boarding school is a superstore and is stocking the shelves on the shopping aisle labeled: "Academic Passion and Curiosity," then it is looking to stock many different types of brands, offering many different qualities. Simply put, boarding schools are looking for students who have different things to offer.

What are your academic passions? Are you a student who is curious about history and has read numerous books on the subject? Do you have a particular interest in World War II and can cite historical facts about this period with ease? Or are you a student who is passionate about writing and current affairs and has written articles for their school newspaper or is a member of their school's Model United Nations team for the past few years? Maybe you are the type of student who likes to build things and studies Python and Scratch in your free time and wants to enter robotics competitions. Students who are eager and excited about learning bring that vibrancy and energy to the schools that they attend.

Start Early

Find what you like and then work on developing this interest so that you can demonstrate this to your customer: the boarding school. Don't define yourself only by what you are good at, as this will prevent you from exploring a different subject or activity. Schools are looking for passion, as well as proficiency. A few years before your boarding school application, try to explore other areas of academic interests that either are not available at your school or are beyond the offerings. Students who demonstrate not only academic proficiencies but academic curiosity as part of their brand can distinguish themselves. There are many resources for the student to explore academic passions, whether it be in specialized classes or summer programs, online learning, or independent study projects. Some of the larger boarding schools have several hundred courses from which to choose. Students who demonstrate a love of learning for the sake of learning are a naturally desirable brand for most schools and will likely find a place on the shelf of the "Academic Passions and Curiosity" aisle of the superstore.

X-Factor 2: Athletics

Most boarding schools require students to play sports. Depending on the school, students may be required to play a different sport for each of the three seasons: Fall, Winter, and Spring. Other schools may require that a student participate in one team sport per year and offer the availability to fulfill the afternoon activity commitment in other ways, such as through intramural sports, drama, or different types of extra-curricular activities. Almost unanimously, students will be required by their customer — the boarding school — to play some sport at some time in their boarding school career. In defining your brand, it is beneficial to have some sports-related skills.

Turning down the sports equipment aisle of our superstore, there are many different types of sporting equipment and many different price points reflecting certain levels of products. You can find a tennis racquet for an intermediate player or one that is top of the line for an expert. There is equipment for soccer players, swimmers and an array of other sports. For the students who need the top-of-the-line tennis racquet, tennis can be an X-Factor to have as part of their brand. These students play tennis regularly on their school team and likely participate on a club team outside of school. They will probably attend summer camps to help improve their skills, join competitions, and have some accomplishments in tennis competitions to share on their applications. Perhaps other students have some brief experience with soccer and like to run but have never played on a formal soccer team. While the superstore has top of the line soccer equipment to sell, these students only need the middle of the line brand because soccer is not a very strong X-Factor.

Fit is important when developing the X-Factor of Athletics. Superstores can't possibly carry every brand. Students should make sure that they are approaching the right superstores with their brand. If a student's brand is made up of the X-Factor of swimming and the school does not have a competitive swim program, the brand mission is not consistent with the superstore mission, and the brand will not hold a lot of value.

Common sports at American boarding schools*

Fall Season	Winter Season	Spring Season
Soccer (Girls and Boys)	Basketball (Girls and Boys)	Lacrosse (Girls and Boys)
Cross Country Running (Girls and Boys)	Swimming (Girls and Boys)	Baseball (Boys)
Field Hockey (Girls)	Wrestling (Boys)	Softball (Girls)
Waterpolo (Girls and Boys)	Skiing (Girls and Boys)	Track and Field (Girls and Boys)
Football (Boys)	Indoor Track (Girls and Boys)	Crew (Boys and Girls)
Volleyball (Girls)	Ice Hockey (Girls and Boys)	Tennis (Girls and Boys)
	Squash (Girls and Boys)	Golf (Boys and Girls)

*Offerings vary by school and seasons can vary by location.

Some students have minimal athletic features to their brand, or they have spent time in athletic training for sports that are not readily available in US boarding schools. Most boarding schools appreciate some athletics as part of your brand. Starting early to gain an understanding of how developing American boarding school sports can help students to build a stronger brand. There are two ways to look at sports as X-Factors or part of your brand.

1. The first way is that sports are your brand, a "student-athlete." This type of student will have demonstrated that they are capable of contributing to the classroom and outside of the classroom. These students have a proven, extraordinary ability in a particular sport that is available and desirable at a specific boarding school. Their skills in the area of sports makes these students different from the other students who have sports as part of their brand.

2. Since sports are mandatory in most boarding schools, schools offer many different levels or opportunities to play sports. The most competitive level for each sport is usually called varsity. A varsity team is composed primarily of impact athletes whose X-Factor is sports. Schools also have junior varsity or J.V. teams in most sports. J.V. teams are usually comprised of younger, developing athletes who are training for a spot on the varsity team in later years. Schools also have less competitive teams below J.V. Regardless of the competition level, the teams usually practice 5-6 times a week and have competitions against other boarding school teams at their level.

Start Early

Students should start by researching the sports that are offered by some of the boarding schools they are targeting. Students should try some different types of sports and find a few that they enjoy and think they would like to explore further and perhaps play at boarding school. Once identified, students should work on developing the skills and training for a particular sport and try to enter competitions. It is desirable if at least one sport a student participates in requires cooperation and teamwork to succeed. For example, soccer is a team sport, whereas tennis is more of an individualized sport.

Many schools outside of the US do not offer sports as part of their curriculum. Therefore, students should try to find resources outside of school to gain some sports training. Sports, like swimming, soccer and tennis are readily available in many parts of the world. However, in recent years, there has been increased availability worldwide of some of the other sports that are offered in boarding schools, like crew, lacrosse, and, for girls only, field hockey. Additionally, if students are not able to locate resources close to home, there are many summer camps in the US and elsewhere that offer opportunities for training in most sports.

X-Factor 3: Arts

"The Arts" is a broad term at boarding schools. Most schools require students to take at least one class in the arts. For the purpose of our discussion, Arts refers to everything from visual arts, drama, and music to filmmaking, furniture design, and fashion design. There are many talented artists and musicians that apply to boarding schools. If you are one of these students, then your art talent is an X-Factor and part of your brand. You will want to prepare audition videos or art portfolios to be evaluated by the faculty heads of these departments at the boarding schools. Like the talented athletes, students with a certain proficiency in arts can have an impact at many schools. However, like the need to fill lower, less competitive athletics teams, there are many opportunities for somewhat lesser artists and musicians to be able to include art in their brand. That being said, some students spend a lot of time pursuing low impact art activities that are either not common at US boarding schools or do not have a significant impact. For example, many students may play the piano at a very advanced level and find that, at boarding school, there are few opportunities to contribute to the school community, outside of private piano lessons.

Start Early

It is best if your brand has some form of artistic expression that aligns with opportunities at the schools you are considering. When starting early, take a look at the types of art offerings at boarding schools and see if any of them interest you. It could be that you think you will like oil painting, or you enjoy building and could get involved in theatre set design. Perhaps you are debating whether or not to continue to take piano lessons or switch to the viola. Starting to plan early will give you the time to develop skills and talents in the areas that can have a meaningful impact on your boarding school application and experience.

X-Factor 4: Clubs/School Involvement

Clubs are organizations within each boarding school that represent different affinity groups. These groups are made up of students who share a similar interest or passion. Clubs are a great way to explore interests that are not offered directly by the school in the classroom or on the athletic field. Involvement in clubs at your current school is also a great way to build your brand. Participating in clubs at your current school allows you to demonstrate an interest in a specific area, showcase your collaborative skills, and sometimes provide valuable opportunities for leadership.

Start Early

Past school involvement is a good indicator of future school involvement. Students who have demonstrated participation in school clubs and activities at their current schools are most likely to continue to be active at boarding school. Schools vary in club offerings, and some schools do not have many options for middle-year students. It is recommended that students explore which clubs are available to them at their current school and get involved in a few. For other students who either do not have clubs available at school or are looking to get involved in exploring an area that is not offered, finding local resources can help a student to explore potential new X-Factors.

X-Factor 5: Community Service

Giving back to the community is part of most boarding schools' missions, and some schools require students to participate in community service events on- and off-campus. Since most students applying to boarding schools are between 12-15 years of age, admission officers know that the opportunities for significant independent community service are limited due to the students' age and maturity.

Knowing that community service can be a significant asset to a student's brand, shortly before applying to boarding school, many students look for ways to participate in some form of community service. They volunteer in "one-off" community service events so that they can mainly "check the box" for having engaged in some community outreach. Parents can orchestrate elaborate community service opportunities while other students may exaggerate their involvement.

Community service should only be part of a student's brand if it is genuine. Students need to be able to talk about their experiences, and the only way to talk about something in any meaningful way is to have a genuine interest in it and have participated in the experience. When planning to attend community service types of activities, keep in mind a student's interests and try to expose them to ways of helping others that are aligned with those interests. For example, a student who is very good at computer science may try to reach out to the elderly in their community to provide computer literacy to this group. A student who has an interest in music may volunteer to play their instrument at a charity event or teach under-resourced kids how to play an instrument.

Start Early

Consistent involvement with a particular community service project is more desirable than participation in a "one-off" event. Students should try to find and involve themselves in opportunities that evolve, allowing them to learn about the organization and develop relationships with the people they are trying to help. Consistent involvement in community service can also provide an opportunity for students to develop leadership skills. The most meaningful community service projects are ones that impact the student's perspective on a particular issue or segment of their community. Firstly, it takes time to find these types of organizations that are willing to allow younger students to volunteer. Secondly, for this activity to be impactful, students need to spend the time participating in the activity.

X-Factor 6: Leadership

Leadership opportunities for middle school students are usually very limited. There are very few opportunities available, or those that are available are designated for older students. At school, students should try their best to find available leadership opportunities. They should find out the criteria for selection and target this as a goal. Leadership at school can range from Class President to Classroom Representative to a sports team captain. Many younger students realize that they need to go outside of the school to be able to find leadership opportunities. Some students can parlay their community service experiences into demonstrated leadership opportunities by organizing events or getting other students involved in the organization. Students who demonstrate leadership skills as part of their brand can be attractive at many schools.

Not everyone can be, should be or has to be a leader. There have to be some students who are the "followers" working towards common goals. Demonstrated participation in various activities and clubs is also a valued brand component.

Start Early

For the most part, leadership skills are honed with practice. Students should try to find small areas to practice leadership skills at an early age. Confidence is an important component of leadership. Consider speaking lessons and participating in activities that require public speaking, like debate. Look for ways to build confidence through achievements. Parents should not discount failure, as it is a way to build resilience and confidence. Many international students have demonstrated their ability to overcome challenges in changing schools, adjusting to new cultures, and more. These types of experiences are invaluable for building confidence. Other students may have not been exposed to any significant challenges or hardships in their lives. Parents may want to consider exposing their son or daughter to experiences that are outside of his or her comfort zones to build confidence, which is vital in developing leadership skills.

X-Factor 7: Awards and Recognitions

There is a section on every boarding school application to designate awards, recognitions, and achievements received by the applicant. Space is usually limited, and most schools are only interested in learning about a student's accomplishments over the past few years. Students and their parents should resist the temptation to include every award or certificate earned from an early age. First, including more than is asked for is an indicator that you have not followed the guidelines that are clearly designated on most applications.

Secondly, providing information that is not timely or relevant creates "noise" and distracts from your brand when an admission officer is reading your application.

List achievements in order of significance, chronological order, or groupings based on the subject matter. For example, a student who is a talented artist may want to list all of their art achievements together to powerfully convey art is an X-Factor or a significant part of their brand. Others may wish to list recognitions in order of significance. For example, receiving academic honors, or a national athletic title would be best to list first, over smaller achievements such as winning 5th place in a local robotics competition.

Start Early

It takes time to develop the proficiencies in areas to warrant recognitions. As early as possible, find out what the criteria are for awards and set that as your goal. Starting to explore awards criteria early will allow you to make the strides necessary to reach your goals. Identify a few areas of interest and strength (again ones that are sold in the superstore!). Work to develop skills in these areas and enter some competitions. Start small at first to better increase the likelihood that you may build confidence and receive recognition. For example, start with a local writing or debate competition to build later participation in an international competition like World Scholar's Cup or entering international literary competitions.

The exercise of putting together a Curriculum Vitae or resume can be beneficial. Remember that participation counts and keep an ongoing list of your achievements so that when it comes time to complete your boarding school applications, you can easily reference this document and prioritize your awards and recognitions.

Think Like a Customer

In Creating an Authentic Message, students are encouraged to think like a customer. Why do you buy a particular product? What makes you choose a certain brand? A successful brand will find out what a customer wants and try to make the brand fit that need.

Thinking like a customer will help you to align your strengths and interests with those of a particular boarding school. Understanding what your customer is looking for is going to help you know how to give them what they want. Going back to the superstore analogy, boarding schools are the customers to which you are looking to sell your brand. They are looking to fill their shelves with a variety of products and brands. Your job is to understand how **You** fit on their shelves.

Remember the three fundamental questions posed earlier, answering them will help you think like your customer:

- ***Why do I want to go to a US boarding school?***
- ***What goals do I hope to achieve at a US boarding school?***
- ***What are my brand's X-Factors?***

Why do I want to go to a US boarding school?

Leaving home as a young teenager to live in another country and attend boarding school should not be taken on lightly. There are many challenges that students face, including academic challenges and residential challenges, from tasks like managing their schedule to homesickness. Students who are invested in the process tend to do better than those that are not. A definite part of a student's message of their brand should be defining why they want to go to boarding school. Whose decision is it, and how did they come to this decision? Perhaps your parents or siblings went to boarding school, and your attending follows a family tradition. For other students, US boarding schools may present academic offerings and programs that are not available in their current school. Others feel that the teaching style of US schools' better suits their learning style. Whatever the reason, students need to have a thought-out and articulated reason why they want to go to boarding school. Your customer wants to make sure that you are in their superstore for the right reasons.

What goals do I hope to achieve at boarding school?

While students are not expected to be able to articulate their goals within the context of a college major or career path, they are expected to have a sense of their goals and aspirations in attending boarding school. Many students and parents are inclined to articulate their purpose of attending a US boarding school to better prepare the student for US college and university. While the admission committee clearly understands this, ultimately, when asking this question, boarding schools are probing for more intrinsic motivators from students. Boarding schools are looking for students whose goals match those of the articulated missions and programs available at that particular school. For example, one student may want to study advanced biochemistry while another wants to learn to be a better writer or debater, another desires to develop their acting skills, and yet another to learn a new instrument.

What are my brand's X-Factors?

All students need first to demonstrate that they are academically prepared to participate in the rigorous academics at boarding school. Once the academic fit has been determined, the key to admission is **differentiation**. What are your X-Factors? How impactful are they to the boarding school, and how can you demonstrate them to your customer? Different schools value different X-Factors, and the needs of the schools can vary based on the demographics of a specific cohort or the mission of the school. For example, a school that has invested millions of dollars in a new robotics program will value those students with robotics as an X-Factor to build their program. Other schools may have specialized programs and are looking for students to fill those programs, whether it be sports, engineering, or music, for example.

It is important to know about the programs and cultures of different boarding schools so that you can understand how to represent your brand and your X-Factors to that particular school. Does one school seem to place more importance on sports? If so, how do you showcase your athletic experiences? Does another school seem to place a high value on community service and giving back to the community? If so, how do you develop or highlight this part of your profile? Doing the research and visiting schools will help you to identify if and how your brand, your X-Factors match up to a particular school.

Create Brand Loyalty

Creating brand loyalty is crucial to the success of any business. Once you have established your brand, students need to build the network around them to support the promotion of their brand. In the boarding school process, bra loyalty is achieved by finding people who believe in your brand an willing to promote your brand to the boarding school. These would loyal customers or supporters.

Parents

Your most loyal supporters are your parents. You likely have already built the customer loyalty of your parents if you are applying to boarding schools. Your parents have supported you thus far in providing resources and time to help you develop your X-Factors. Parents also support their students by assisting them in arranging school visits and accompanying them to the US (later Steps will discuss how parents can support their children in the parent interview and parent statement). Your parents are your champions and will do everything within their power to help promote your brand to the customer.

Teachers

Teachers are an excellent source to build brand loyalty because they have direct experience with your brand. Building brand loyalty among teachers is easy for some students. They are engaged students who have meaningful contributions to the classroom each and every day. They demonstrate academic curiosity and produce excellent work. When it comes time to ask teachers for a recommendation, their teachers write glowing reports about the student's brand and their experiences with the student. It is like reading a customer review of a product that you are thinking about buying. Think about recommendations as customer reviews. A good review, or recommendation, can provide an admission officer with information about whether or not your product is something they are interested in putting on their shelf. Unlike your parents, teachers are more objective, and admission officers place great value teacher recommendations. Building the type of brand support that produces ing teacher recommendations can take time.

nt place to start to find out how strong your brand is with your

teachers is through parent-teacher conferences and in teacher comm... grade reports. Look for feedback on how you can build their support. Does the teacher see specific weaknesses in your brand, like class participation or not carefully checking over your work? Are there some behaviors that you exhibit that make it very difficult for the teacher to support your brand, like misbehaving in class or turning in assignments late? All of these factors can be remedied with time if you are aware of them and can rebuild your relationships through excellent customer service.

Coaches

Coaches are another source of excellent brand loyalty. Many coaches have had years of experience with your brand. Coaches can speak to how your brand has grown and improved over the years. As your coach, they can also identify how well your brand stacks up when it is challenged and tested. Having experience with a brand over time is an important piece of information for your customers to know as they consider the stability and resilience of your brand power.

Extra-Curricular Activities Coordinators

Students should be involved in extra-curricular activities, both in school and outside of school. Hopefully, if you have given your X-Factors some time to develop, you have several loyal brand supporters in the area of extra-curricular activities. These types of loyal customers can share how your product performs in their specific field of reference or interest. A **Special Interest Recommendation** from a music teacher or art teacher can provide insight into a student's skills in this area. Special Interest Recommendations also allow another unbiased and valued professional to assert their opinion of a student's performance compared to other students of similar age.

Community Organizations

Community organizations like churches or non-profits that a student is involved in can be another avenue to promote your brand to your customers. Often these loyal supporters can provide a different kind of testimonial about your product. A **Personal Recommendation** from the organizer of a charity at which a student volunteers or the leader of the student youth group can give support about how your brand interacts with and supports others.

BRANDING PART 2: PROMOTING YOUR BRAND

You have created an authentic message in identifying your brand and built a base of strong customer loyalty for your brand. Well done!

Promoting your brand enables you to show the benefits of your brand and what kind of service your brand can provide to the customer. Now, how are you going to start **Promoting Your Brand** to boarding schools?

Develop a Marketing Strategy

Many intelligent and talented students apply to boarding school each year. Hopefully, you have demonstrated your academic readiness and created an authentic message to be considered one of these students. However, not all students do an excellent job of promoting their brand to the boarding schools. Poor promotion results in an application that doesn't resonate, or "pop," with admissions. Without a good marketing strategy, students may find it challenging to convey to the admission committee how and what they can contribute to their school. Nothing is compelling the committee to accept this student. Developing a plan on how to market yourself to boarding school is key to making your application stand out among the others. Remember, at this stage, it is all about **differentiation**.

Evaluate your X-Factors and **Develop a Marketing Strategy**. Do you have a few key attributes or X-Factors that are likely of high impact? Do you have several X-Factors that allow you to spread your impact across many different areas of a boarding school's interests?

Develop a Social Media Presence

Boarding schools are certainly using social media effectively to promote their brands to families. In our tech-savvy world, most of us are using one form of social media or another, so why not have social media as part of your marketing strategy to promote your brand? Depending on their X-Factors, students can utilize social media as part of their marketing plan. It may be posting a photo of yourself in your school drama production or publishing a video of you playing your instrument or sport on YouTube. Other students may have their own YouTube channel or a blog that they utilize to share their brand regularly. Social media is an effective way for students to promote their brand.

Keep in mind that your **Social Media Presence** is a representation of your brand. Use social media to your advantage but remember to keep it clean. Boarding schools can easily see anything, and everything students post on social media.

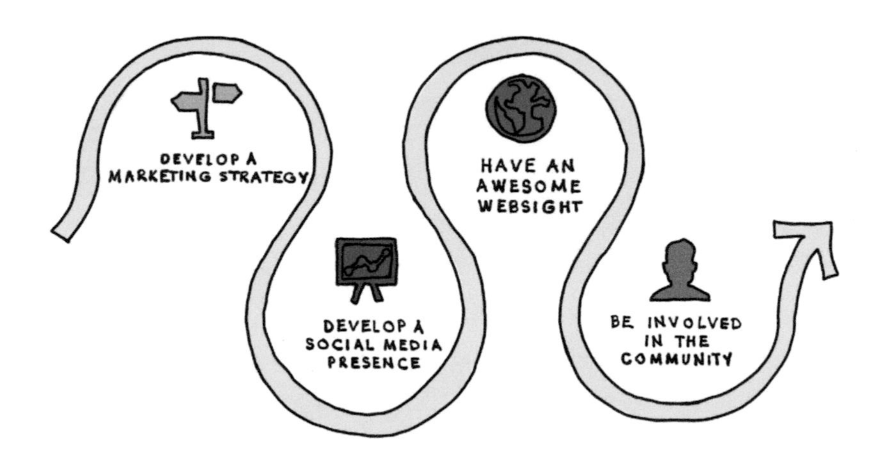

Have an Awesome Website

Students should evaluate how they want to communicate their brand. Students are given the opportunity in the application process to provide multi-media links to supplement their application materials. For some brands, creating a website can be an effective marketing strategy. Students who have visual materials, such as art portfolios, may find that an Awesome Website best helps them communicate their brand attributes. Other students can use a website as a receptacle for information that they want to highlight or satisfy the need to share multiple pieces of related information in one place. For example, if writing is an X-Factor, a student can create a website to communicate their writing talents by including creative writing, poems, or published writing all within one site. There are many free website builders that students can use. Some students have the skills to design and build their own websites, whereby the website development itself can be an X-Factor. If a student creates their website, they should be sure to let the boarding schools know that they have done this.

Be Involved in the Community

It is difficult to promote your brand if nobody knows about you. Get out in your community and share your brand. Meet people and be open to feedback. Testing out your brand is the best way to know how your brand performs and how to market it. If your brand involves athletics, make sure that you try some competitions or matches. If your X-Factor includes music, participate in the orchestra at your school or community youth orchestra. Writers can enter writing contests or submit articles to their school's newspaper. Artists should be learning to communicate with others about their artistic visions and seeking feedback from other artists, peers, and professionals on how to improve their art or reach beyond their current experiences.

1.3 BUILDING RELATIONSHIPS

When the goal is brand recognition, building relationships can be beneficial and often strategic. Building relationships is equally important for students wanting to build their brand to a boarding school. When it comes to the boarding school process, **Building Relationships** is a two-way street. It takes time for boarding schools to get to know students to determine if they are a good fit for a school. It also takes time for students to get to know and experience particular schools to see if they are a good fit for them. Meaningful relationships take time and attention to build.

PRELIMINARY VISITS

Most families decide on which boarding school to attend based on only a 2-hour school visit and whatever information that they can gather online or from their friends. It makes more sense for both parties to get to know each other over time. **Preliminary Visits** are a fantastic place for students and boarding schools to start getting to know each other and build the relationship.

Early in the process, students should visit a variety of schools, some large, some small. Perhaps visit a rural school and another school that is in a more suburban setting. Students should research programs that relate to a specific X-Factor they have and seek out schools that are well known in that area.

Visiting some highly competitive schools and some schools that are slightly less competitive will help families to understand the depth and breadth of offerings available at most US boarding schools.

A preliminary/early school visit usually consists of a tour and does not involve a formal meeting with an admission officer. Generally speaking, most schools will not welcome younger students to visit during the school's peak times when current year students are visiting, such as October and November. Some schools will also not offer a guided tour to students who are not in the current year application process but will welcome families to campus for a self-guided tour. Don't be turned off by a school that does not offer preliminary visits during busy times, such as in the fall. Keep in mind that when your child is a current year applicant, you will appreciate the school's focus at that time.

Good times for a preliminary visit may be throughout the summer, during Chinese New Year, in the late spring after April 10. Summer visits will provide a different experience than a visit during the academic school year. The most obvious difference is that enrolled students will not be on campus. Since the students are not on campus in the summer, if a school can provide a student tour guide, they may be a local day student who is helping out at the school for the summer by acting as a tour guide. Many boarding schools also offer summer programs, so

that the students that are on campus may not be representative of the actual student body.

For students and families who have no prior US boarding school experience or exposure, the US boarding school process can be daunting and overwhelming. The experience of stepping foot on a school campus early on in the process helps students to start to imagine themselves at a particular school. With the limitations mentioned above, summer visits still are a great introduction to schools. Summer visits can also help a reluctant applicant become more confident. For a student who is not 100% on board with this process, visiting early may help to provide a clearer understanding of what a boarding school experience would be like, and help further inform their decision to apply. Preliminary visits can also help students who need added motivation and incentive to prepare themselves for the difficult and time-consuming process of applying to boarding school.

To plan a preliminary visit, students must complete the online inquiry process registering the student in the school's database before setting up a preliminary visit. The online inquiry process is found on the school's admission page and involves providing demographic and school information as well as completing interest and activities questionnaires. It is also important to register in the school's database to receive future invitations such as local receptions and event announcements. Once a student is in the database, they will be able to contact the school to request a preliminary visit.

Preliminary visits are not yet time to settle on a final school list. Preliminary visits serve to start the process of getting to know schools and schools getting to know students. There is plenty of time to narrow down the school list.

Local Receptions

Local receptions are another excellent venue for international students to learn about schools and start to build relationships with schools before or without necessarily traveling to the US for a preliminary visit.

Many boarding schools host receptions in cities around the world. Most boarding schools will post their international travel schedule on their admission page of the website. Based on past admission office travel schedules, boarding schools participate in these types of international receptions most often in the fall between September and early December. Students must be in the database to receive the invite and attend a reception in their home city. In larger cities, receptions can reach capacity very quickly. Early registration is critical to ensure your place.

The purpose of **Local Receptions** is twofold. Schools travel to many cities in the US and around the world to continue to build relationships with current local students and their families and to promote their schools to future students. Most boarding school admission officers will also carve out some time during their visit to conduct interviews with current year applicants.

What to expect at a local reception? Local receptions are most often hosted in collaboration with the local parent/alumni group in a particular city. Receptions are usually held at a hotel or function center. Some of the smaller receptions may be held at the host's home. The receptions are usually attended by several representatives from a school such as a representative from admissions, often accompanied by a representative from the school's advancement office and perhaps a college counselor or a summer program director. Some Head of Schools will also attend the receptions.

The format of these receptions is quite similar. Most receptions will contain a presentation and video about the school followed by time for prospective students and families to ask questions. Some receptions may have a panel of current students and alumni who share their experiences at the school and beyond. Other schools may also have parents give testimonials. Receptions are vehicles for the school's administrators to present their school to prospective students and families. Like you, they are trying to establish their brand in a particular market.

Many students head off to local receptions hoping to get time to promote their brand to the faculty and staff of the school. Many receptions, particularly the most competitive schools and in the larger or capital cities of the world, can have up to several hundred students attending the event. Students should undoubtedly try to interact with the admission officer if possible. However, students should not expect or plan to have a conversation that takes more than a minute or two of the admission officer's time. All students should be prepared to introduce themselves. Most admission officers will ask students a few questions about themselves, such as, "What school are you currently attending?" "What types of subjects are you most interested in?" "What types of extra-curricular interests do you have?" "How did you learn about my school?"

During the reception's Q and A period, admission officers also encourage students to ask questions. Boarding schools like students who are willing to stand up in a crowd of people and ask a question. Beware though about the question(s) that you ask. Keep in mind, if a student asks a question that has already been answered, that student will come across as not having paid attention. Some students ask very specific questions at receptions that pertain only to themselves or only a few individuals. Asking a personal question is also not appropriate in a public forum. In both of these scenarios, it is better for the student to say nothing rather than call attention to themselves in a negative manner. Students should prepare a few general questions about the school and ask a question only if the question has not been answered previously.

Many families ask if there is an advantage to interview in their home city. For students who may have had a preliminary visit and have previous direct experience with a school, interviewing locally can be an option. These students have already started to establish their brand and are familiar with their customer. An official tour and interview during the current application year are always the best scenario for success. However, if it is not possible or practical for families to travel to the US, interviewing with admissions when they are in a student's home city is an option. Local interviews are usually conducted at the hotel lobby or meeting room where the admission officer resides. Students should plan to schedule local interviews well in advance. Admission officers' local interview schedules book quickly, and they adhere to a strict timeline for the interview.

STEP 02 PLAN WELL

2.1 – SET YOUR TARGET
2.2 - GETTING READY TO VISIT SCHOOLS
2.3 - STUDENT AND PARENT SCHOOL VISIT

The importance of starting the boarding school process early has been identified. Early preparation gives students time to develop their academic readiness, develop their X-Factors, build their brand, develop a communication plan, and build customer loyalty. By this time, students should have a clearer idea of their strengths and begun to develop them further and work on improving their weaknesses. Starting early also provides time for students and boarding schools to become acquainted with each other. It's time to start Step 2: Plan Well.

2.1 SET YOUR TARGET

Compiling a list of target schools can be a complicated process. Most families are unaware of the breadth and depth of boarding school offerings beyond the top 10 that they consider famous or are listed in the published rankings. Unaware of the need for a well-thought-out list and lacking a clear understanding of how competitive the process is, some families apply to only the most competitive schools. Often families who approach their school list in this manner may find that after all of the hard work of visiting and applying to boarding schools, no superstore is interested in putting their product on the shelves. Having no offers from any boarding school is a difficult situation to be left with. It is essential for any brand to know who its target market is and to plan accordingly.

Identifying a list of your target schools involves knowing yourself, what you are looking for in a school, and what schools are looking for in students.

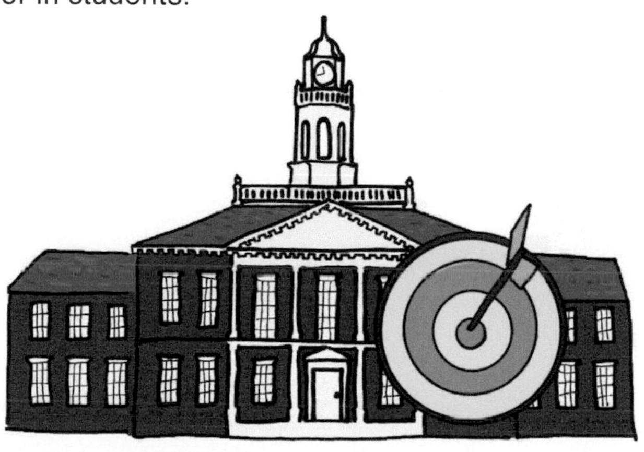

UNDERSTANDING YOUR STRENGTHS AND WEAKNESSES

Here is where knowing your brand and your customer is so critical. School "fit" is of the utmost importance to your customer, so you must align yourself with the expectations of the boarding schools to which you are applying. **Understanding Your Strengths and Weaknesses** is essential for you to target a list of appropriate schools that may be interested in your brand.

Students are recommended to examine themselves in each category identified in Step 1, and below, and realistically evaluate themselves in these key areas. You can get a better grasp of your strengths and weaknesses by asking yourself the questions listed in each category below.

Academic Readiness

English Proficiency

A student's level of English proficiency must be aligned with the expectations of the schools they are considering.

- ***Are my English skills such that I will not require any additional English support?***
- ***How does my TOEFL score match up to the requirements of the schools to which I am applying?***

For example, if you are applying to only top schools that likely require a TOEFL score of 105 or above and your TOEFL score is 85, it is improbable that you will get admitted, and your brand will be collecting dust in the proverbial warehouse. Instead, if you identify schools where your score of 85 is within range of their TOEFL requirements and a few where it is slightly above, your English proficiency is more aligned, and these schools are more likely potential customers and appropriate schools for you to have on your school list.

Academic Curiosity

Students who demonstrate academic curiosity as part of the application process are more attractive to boarding schools than those who do not have an academic spark.

- ***How have I communicated academic curiosity to the schools?***
- ***Did I pursue some areas of study independently or take a summer course in an area of interest?***
- ***If you pursued a project to demonstrate your interest in an area, how impactful is this project to the particular schools on your list?***
- ***Do I show the 4 C's of Critical Thinking, Creativity, Collaboration and Communication?***

Grades

Grades represent the most critical factor in determining academic readiness. It makes excellent sense that a student who is doing well academically at their current school will be more likely to do well in academics at boarding school than one who is not. Grades represent longer-term data for schools and are not as influenced by test prep, as standardized testing can be.

- ***How do my grades stack up? Are they among the best in my class, or is my academic performance more representative of the mid-range of my peers?***
- ***Are there some subjects that I naturally excel at over others?***
- ***Are the boarding schools familiar with my school and its curriculum?***

SSAT

High SSAT scores will never be the deciding factor that gets a student admitted to a school, but low SSAT scores can certainly prevent students from being accepted.

- ***How do my SSAT scores match up to the expectations of the schools on my list?***
- ***Are my SSAT scores consistent with my grades and level of English proficiency?***
- ***Is my Critical Reading score high enough to give the school confidence that I not only have the reading skills necessary to be successful, but also the ability to analyze and infer from the text that I read?***
- ***Are the writing skills that I demonstrated in the SSAT writing sample aligned with what the schools on my list expect?***
- ***Is the level of writing skills in my SSAT commensurate with my application essays?***

Teacher Recommendations

Teacher Recommendations are an important evaluation tool for the schools. Boarding schools value other educators' perspectives on working with you and your contributions in the classroom and school.

- ***Am I expecting my teacher recommendations to be favorable?***
- ***Are there certain subjects that are weaker than others?***
- ***Am I not entirely confident in what my teachers will say about me?***

Athletics

Hopefully, you followed the advice in Step 1.2 and spent some time building your brand so you can contribute athletically to the boarding schools you are applying to.

- ***How would my participation in the sport impact the school?***
- ***Would my sports contribution be competitive or potentially high impact?***

Arts

- ***Would I consider my impact to be significant?***
- ***Will I have a portfolio, website, or multi-media link to support my interests?***

Clubs/School Involvement

- *Would I be considered an active member of my school community?*
- *What types of clubs and school involvement have I engaged in for the last few years?*
- *Are my interests aligned with any clubs offered at the boarding schools I am considering?*

Community Service

- *Have I been able to make a significant impact in my community?*
- *How did I demonstrate and communicate this?*
- *Is my community service genuine and aligned with my interests?*
- *Did I have any leadership roles?*

Leadership

- *What types of leadership have I undertaken inside and outside of school?*
- *What kinds of skills have these leadership experiences provided to me?*

Awards and Recognitions

- ***What types of awards and recognitions are going to be part of my application?***
- ***When did I achieve them and, are they significant?***
- ***Do they position me to be considered high impact at a particular boarding school?***

Look at each category and realistically evaluate yourself in these areas to align yourself with a list of boarding schools that best fit your skills and personality. Let's go back to the fundamental questions started with in Step 1.2: Build Your Brand. Let's see how you have done:

- ***What is special about me?***
- ***How am I going to be remembered?***
- ***What impact can I have at a school?***

WHAT SCHOOL IS BEST FOR ME?

You now understand your strengths and weaknesses. You have worked to establish your brand, which includes the many factors mentioned in the previous section. Some elements are stronger than others, but your brand should now have more to offer.

After your self-evaluation, be aware of how your academics, interests, and activities align with a particular school. Would you do best in a school that offers English language support? A small school or a larger one? If you are interested in swimming and intend to participate on the swim team at boarding school, make sure the schools that you are applying to have a swim program. Beyond merely the availability of a program, particularly as it relates to sports, it is prudent to make sure that your ability aligns with the sports program. An athletic program that is too competitive will likely result in the student being frustrated because they are not getting enough opportunity to participate. A program that is not competitive enough may prevent the student from further developing their skills in anticipation of, perhaps, competing at the collegiate level.

Keep in mind that even if you have a demonstrated ability in an X-Factor, if the X-Factor is not reflected in the school's programs, your X-Factor holds NO value to that school. If you are a championship bowler, for example, there likely is no added value of this accomplishment because most boarding schools do not offer bowling as a competitive sport.

School fit is a two-way street. Not all schools will be interested in your particular brand, and you will not be interested in your brand being represented at all schools. How do you start to find schools that offer not only what your brand represents, but may be willing to put your brand on their shelves? A strategic approach to this question should be considered. Is it better to select schools that will place your brand on a prominent shelf in their store, front and center, or do you want a school where your brand might be in the back, and challenging for the customer to see? The best place to start making a school list is to do some research on different schools.

Research

Researching the following variables will help narrow your school list. Most of the information about these variables is readily available online and therefore not discussed in this book. What is important is to research the variables.

Geographic Location

Boarding schools are located throughout the US with the densest concentration on both the East Coast and the West Coast. Some families have a distinct preference for attending a boarding school in a particular state or region. For some, this is due to having family close by to the location. Others are looking for proximity to a larger city for ease of transportation or access to cultural events. Do you want a boarding school located in a rural area, or would you be comfortable with a location that is closer to a city? Geographic factors can help to narrow your list immediately.

Size

There is variability in the size of the student bodies and campuses at boarding schools. Hopefully, you know what kind of an environment you would tend to thrive in. Do you like a smaller school where there is more contact with staff and faculty? Do you think you would excel in a larger school that can provide an extensive curriculum similar to a small college? Do you like a medium-sized school that is big enough that you feel you will have opportunities that a larger school would offer, but small enough that the sense of community will likely be very strong?

Percentage of Boarding

There are very few 100% boarding schools where all students and faculty live on campus. Most boarding schools have a mix of students who live at the school and local students who attend the school during the day but go home each evening. For some students, the ratio of boarding to day students is an important factor in determining their school fit.

Coed or Single-Sex

Most boarding schools are co-educational, with males and females comprising the student body. A single-sex student body where the school contains only male or only female students can be an excellent fit for some students. Students are encouraged to do some research and see if a single-sex school is for them.

School Culture / Saturday Classes / Dress Code

Some schools require students to attend classes on Saturdays. Some schools have a very formal dress code where boys are required to wear a jacket and a tie to class, and girls are expected to dress in a business casual style. Other schools have little to no dress code, and students wear t-shirts and jeans. Determining what you want your experience to be in the areas of Saturday classes and dress code will further define your list.

Program Offerings

An essential factor in deciding which schools to apply is program offerings. Which schools offer the programs you are looking for? Are there particular schools that have established programs to support your X-Factors? Do they offer enough opportunities for you to explore programs and try new things in the future? Are you looking for AP or IB curriculums?

Selectivity

How selective are the schools on your list for you? Is your list made up of only the most selective schools? How likely are you to get an offer from the schools on your list?

A BALANCED SCHOOL LIST

Approaching this process with a balanced list is essential. The number of schools that international students should consider is not a set number and may vary by the grade students are applying to. For most schools, Grade 9 is a highly competitive entry point, but also the entry point that has the most available openings. Some general guidelines for the number of schools that international students may want to consider applying to are as follows.

How many boarding schools should international students apply to?

Grade Applying	Suggested Number of Schools Student Should Apply To
9	8-10
10	10-12
11	Cast an extensive net as many schools do not have many offerings past grade 10.
12	Most schools do not accept applications to Grade 12 unless under special circumstances. Students wishing to apply to Grade 12 are encouraged to reach out to a school to check availability before applying.
Post-Graduate	Will be addressed later in "Special Considerations."

Students should consider a school list with schools of varying selectivity. There are many different terms to classify schools. The terms — Highly Selective / Selective / Slightly Less Selective — are used here. Selectivity is based on an individual's overall profile. What is considered selective for one student may be highly selective for another student. If a student is applying to 8 schools, the following distribution of selectivity is recommended:

2 Highly Selective Schools

4 Selective Schools

2 Slightly Less Selective Schools

How to build a balanced school list?

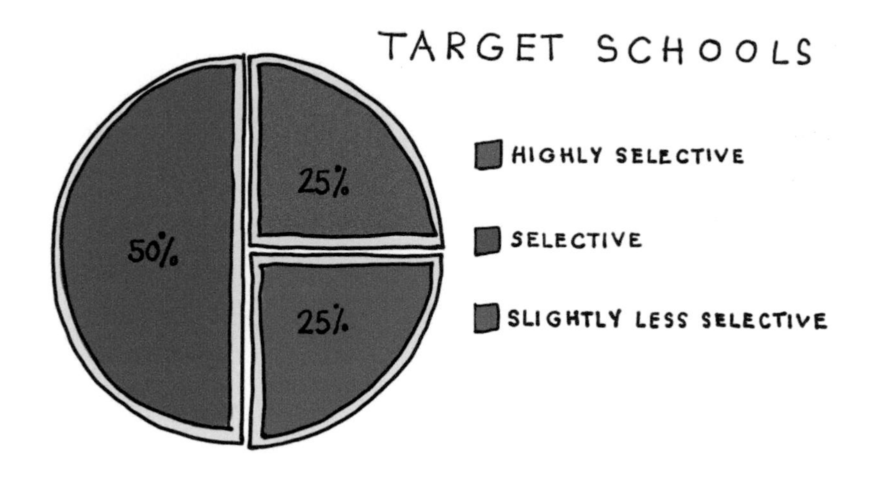

The Big Fish Success Story

When it comes to fit, many parents approach this process with the goal of their son or daughter getting into the best school possible. This approach usually translates to the schools with the highest rankings or those with the most famous brands. Most parents also hope that the boarding school experience will result in a prestigious college or university placement for the son or daughter; most articulating a preference for Ivy League schools.

Understandably parents want the best for their children. Everyone knows the Big Fish in the Small Pond story. This story is relevant to boarding school admission, success in boarding school, and college placement outcomes. Time and time again, students who select a boarding school that is the right fit for them, regardless of ranking, thrive at the school. The boarding school they chose is at the right level, so they can take advantage of all of the opportunities offered at the school. Students who find themselves the "big fish in the small pond" will most likely enjoy excellent college placement results.

It is the big fish that stands out in the process.

When fit is not a primary consideration in school selection, some students can wind up in a boarding school where they have difficulty standing out. This was referenced earlier when discussing applying to a boarding school where your brand is going to be placed on the back shelf, unable to compete with all the others in the front. Students who find themselves admitted to a boarding school where their brand cannot shine can wind up unable to succeed, or at best succeeding marginally. Some students actually "go out of business" and return home. Returning home from boarding school is a devastating outcome on all levels. Students who do not succeed at boarding school suffer the emotional consequences associated with failure. The damage to their self-esteem and confidence can be shattering. The US education system is based on a cumulative Grade Point Average or GPA model which begins with grades from Grade 9. Students who do not thrive risk having a low GPA. Since they are already well into the college process, poor academic performance can negatively impact their college application. Withdrawing from a boarding school before graduating raises red flags in the college process. There is no good outcome for a student who finds themselves at a school that is not a good fit for them.

SPECIAL CONSIDERATIONS

Not all students are alike, so naturally, not all boarding school journeys are simple or straight forward. For example, some students are not quite ready for the process or decide to apply late in the game without adequate time to prepare. These students may want to consider repeating a grade. Some students, though with excellent grades and profile, have low SSAT scores, despite considerable effort and preparation. Others may have a set of circumstances that require specialized services or programs, such as the need for English language support or a learning difference. Older students may be looking to attend boarding school to help with their college application or athletic recruiting processes and are considering attending a post graduate year of boarding school. Younger students may be hoping to join a junior boarding school to have an early experience with a boarding school and thoroughly prepare for the road ahead. Each situation requires Special Consideration and planning. Let's discuss each situation in more depth.

Repeating a Grade

Repeating a grade is a term that is not reflective of the experience. This term refers to a student who enters the same grade from which they are applying. A student repeating Grade 9 at boarding school has completed Grade 9 at their previous school. The term "repeat" is not necessarily the most accurate when you consider subject matter. For boarding school, because students can be placed in classes based on ability, a student repeating 9th grade might find themselves in a calculus class with 10th and 11th graders. When considering whether or not to **Repeat a Grade**, it helps to talk through this decision with a professional, but here are some general guidelines.

What are some of the circumstances where students would consider repeating a grade at a US boarding school?

1. Academic Readiness

Students who are academically behind due to a school transition, learning difference, or other academic difficulties. Students whose academics are not in line with school list expectations.

2. Maturity

This relates back to our discussion in Step 1 about Social and Emotional Readiness. Students who do not have the organizational skills or emotional maturity to handle the demands and stresses of boarding school and being away from home may want to consider repeating a year when they enter boarding school. This also relates to size as it pertains to physical maturity, particularly in boys. Students may not have yet experienced puberty and are very small and immature physically. Some students may benefit from an extra year of growth and maturity before entering a residential environment with much older and more developed peers.

3. Sports Advantage

A student whose X-Factor is sports and has a high likelihood of excelling in the sport and perhaps playing in college, may find that an extra year of maturity and growth can provide physical and emotional advantages.

4. Age

Students who are younger than grade-level peers and who demonstrate academic and maturity issues may benefit from repeating a grade. Being with similar age students can be a better fit emotionally.

5. Absenteeism

Students who missed a lot of school due to illness can benefit from repeating a year.

What are some of the drawbacks of repeating a grade?

1. Age

If a student is already the oldest in their class, repeating their grade can result in them sometimes being two years older than their counterparts. This is accentuated in later grades when older students may be driving, or become restless with the age difference, particularly in their senior year when their classmates and friends from their old school would be college freshman.

2. California Boarding Schools

Students applying to boarding school in California should check with the boarding school athletic eligibility requirements for 12th grade and Post Graduate Students, as there are restrictions that pertain to the number of years students are eligible to participation in sports.

3. Buy-In

Students who are strongly opposed to repeating a grade may be more likely to find reasons not to make it work. Students should be involved in this decision.

Low or Inconsistent SSAT Scores

Low SSAT scores are relevant to how competitive a school list is. What is considered low SSAT for one school may be acceptable at others. SSAT scores are one component of the application and should be considered as one data point in determining fit. However, low SSAT scores will keep students from being considered at some schools. Students with low SSAT scores who are motivated but still unable to improve their scores should try to discover the reason for their low scores. Sometimes simple test taking strategies can improve scores. Conversely, a difference in learning style in preparing for the test may prevent a student from doing their best. All students should consider a school list where their scores are within range of what a particular school is looking for. A student who applies to schools where their SSAT score is well below the school average will likely not be admitted, and their efforts are better spent identifying schools that are a better fit.

Consistency is important when schools are evaluating SSAT and academic readiness. A discrepancy between grades and SSAT scores can raise a red flag by the boarding schools. The following chart provides some perspective on how boarding school admissions could interpret inconsistent SSAT scores.

How boarding school admissions could interpret inconsistent SSAT scores?

Scenarios	Possible Interpretations by Admissions
Low SSAT and Excellent grades	1. The student may not have prepared for the test. 2. The student may not be a proficient test taker. 3. Perhaps there is an undiagnosed learning difference. 4. The academic institution from which the student is coming from may not be very rigorous.
High SSAT scores and Low grades	1. The student drilled for the SSAT, and the score is a result of intense preparation and not a true measure of academic rediness. 2. The student is not motivated at school and won't be motivated at my school. 3. The student is focused on test-taking and may not exhibit the intellectual curiosity characteristics that I am looking for in students.

Academic Support Needs

English Language Learner Support

Students who are not native English speakers can still find a fit boarding school. Many boarding schools offer English support services in the form of a well-defined curriculum where students can take core classes such as English and History within a revised curriculum to strengthen their language abilities. Other schools provide less scaffolding and may offer a specific class for English Language Learners to work on basic English skills such as reading, writing, and grammar, but do not provide revised instruction in core classes.

The question of language ability relates to fit. A TOEFL is usually required for students whose first language is not English and/or whose language of instruction has not been English for the last 2 years. A student who needs further development of their English language skills should apply only to schools that provide English language support. Students who find themselves in learning environments where they are not able to keep up due to their English proficiency will not be well served. A 9th-grade student who is already struggling with English and spending exorbitant hours doing their homework will be buried by grade 10 as the curriculum becomes more difficult. Most schools' English Language Learning programs set a goal for students to enter mainstream instruction so that students are prepared for the university application process.

A listing of boarding schools that offer support to English Language Learners can be found on the Boarding School Review website: https://www.boardingschoolreview.com

Learning Difference Support

Many schools also offer support for students with mild to moderate learning difficulties, with some schools specializing in this area. Most students who have a learning difference will have gone through some type of psychoeducational evaluation. This report describes the student's strengths and weaknesses and provides recommendations for the student's academic and emotional success. Identifying a student's learning support needs is an important part of the boarding school process and targeting schools with the appropriate level of support is critical to the student's success.

The Boarding School Review website also provides a complete directory of schools that offer services for students with learning differences. Transparency is key here. Students and parents should be honest about the student's needs. Transparency ensures that if the school admits the student, they are confident that they will be able to support the student. Schools that provide support are open to students and parents scheduling an appointment with the learning support departments as part of the admission process. Having an initial open dialogue about what the student needs and what the school can provide, allows the family and school to determine fit and sets the tone for future partnership should the student enroll in the school.

Post Graduate Year

A post graduate year is intended for students who have already graduated from Grade 12 from another institution. These students attend a boarding school as a Post Graduate (PG) student for one additional year. If a school accepts PG students, they most often follow the 12-Grade curriculum at boarding schools. The decision for a PG year has some common decision factors as a repeat application to boarding school.

What are some of the reasons why a student would participate in a PG year?

1. Academic Readiness

PG students have already met the degree requirements of graduating from high school, and the extra year of boarding school allows them to make academic choices to enhance their profile for college applications. PG students can fill gaps in their transcripts, take more advanced classes or explore advanced topics.

2. Athletic Readiness

Students enroll in a PG year to strengthen athletic skills in preparation for playing sports in college.

3. Emotional Readiness

For other students, like the repeat application to boarding school, some high school students recognize that they are not ready for the challenges of college or university. Perhaps they have been on the younger side throughout their high school careers compared to their peers and can benefit from a PG year to gain maturity and confidence.

What are some of the challenges for PG students?

1. Left Behind Syndrome

A PG year can be a great experience, but students can also feel left behind as most of their peers from their previous school will have gone off to college. They know their friends have a very different experience as college freshman than a PG student at a US boarding school.

2. Need to Hit the Ground Running

PG students are only at a school for one year, so they need to transition and engage quickly academically, socially, and in extra-curricular activities to take advantage of the reasons for the PG year.

Junior Boarding Schools

Junior boarding schools are similar to secondary boarding schools, but are tailored to the needs of younger students. Most junior boarding schools start their boarding programs at Grades 5 or 6, and some start as early as Grade 3. There are far fewer junior boarding schools than secondary boarding schools and information about most can be found on the Junior Boarding Schools Association website: https://www.jbsa.org.

For the right student, junior boarding school can be a transformative experience and prepare them well for a successful secondary boarding school experience. Junior boarding schools provide similar learning environments, extra-curricular activities and sports as secondary boarding schools. Junior boarding schools offer students developmentally appropriate experiences and opportunities. Students learn early-on how to live independently and are exposed to experiences that help them develop many of the necessary skills for future success in secondary boarding school such as intellectual engagement, teamwork, initiative, resilience, self-control, open-mindedness, and social awareness. The importance of these skills will be addressed later in Step 3.2: Character Skills Snapshot. My experience working with many students to apply to junior boarding school and then continuing working with a great many of them throughout their junior boarding school years to help them apply to secondary boarding school indicates that every student was able to be admitted to their top choice secondary boarding school.

Application Process

The application process for junior boarding school is essentially the same as it is for secondary boarding school except that the deadlines for international students to apply to junior boarding school is usually around mid-December. The most common application used by junior boarding schools is either the SAO or a School Specific Application. Also, depending on the year the student is applying, the student may be required to take either the Lower or Middle Level SSAT. It always best to check each junior boarding school's website for their admission requirements.

Online Resources:

Boarding School Review **https://www.boardingschoolreview.com**

The Association of Boarding Schools **http://www.boardingschools.com**

Ten Schools Association **https://www.tenschools.org/index.cfm**

Junior Boarding School Association **https://www.jbsa.org**

Educational Consultants (For more information, please refer to **Appendix: Educational Consultants**)

Rankings Resources:

Prep Review **https://www.prepreview.com/ranking/us/boarding-school-ranking.php**

2.2 GETTING READY TO VISIT SCHOOLS

You have done your research and identified a balanced school list and taken into account any special considerations that apply. Well done! But there is still more work to do to confirm school fit. Parents and students may feel that they know a boarding school well because they have spent time on its website or because the boarding school has a well-known name and high ranking. Families may have attended a local reception in their home city and were very impressed with the admission presentation and the alumni they encountered. Families can be prematurely convinced that a particular school is a good fit for them and their requirements based on minimal direct experience. This is often a fallacy. Whenever possible, students and their parents should visit any school to which the student is thinking of applying to get the first-hand experience to make an informed decision about fit. Requesting a school visit involves a few steps, some beyond simply reaching out and making an appointment. Before their visit, students are encouraged to reach out to faculty and staff in areas of specific interest to schedule an additional appointment to learn as much as they can and make the most of their time on campus.

THE SCHOOL VISIT TIMELINE

School visits are essential for both the student and the school. Some schools will indicate that the school visit may be optional. With the competitive nature of boarding school admissions, particularly for international students, it is difficult to convey your brand without having stepped onto a particular school's campus. A school visit allows you to meet face to face with admissions staff, students, and faculty. Visiting schools is important for students to be able to have direct experiences so that they can confidently identify their best fit school. Being able to gauge the atmosphere and community of a school's campus is as important to fit as its ranking.

Most students will visit multiple schools during a trip to the US. The coordination of a succinct and efficient route requires early planning. Students should interview during the fall or early winter of their application year. Returning to the example of the current 8th-grader who is planning on applying to boarding school for 9th-grade entry, this student should plan to visit and interview during the fall or early winter of their 8th-grade year. Most boarding schools will begin to book appointments for fall school visits in mid to later summer. Most schools will allow students to visit up until their application deadline or slightly after. Appointments fill very quickly and depending on the time you are planning on visiting, a delay in scheduling may cause an inconvenient travel schedule. If planning is delayed, sometimes interview dates are no longer available. Generally, you should plan to complete all your school visits before the second week of January.

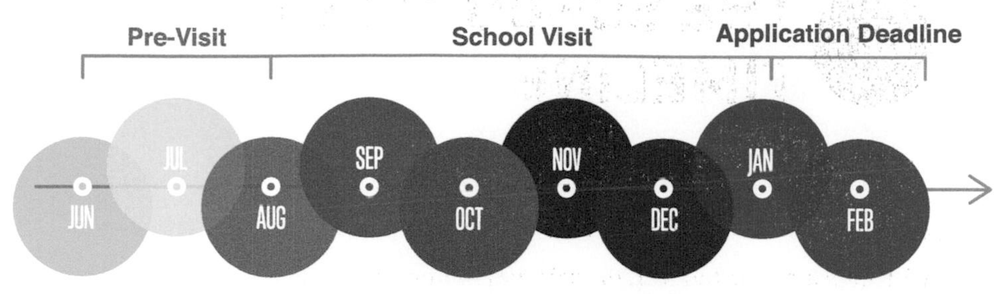

Official School Visit during Application Year Summer

Many parents will inquire about whether they can visit and interview at schools during the summer. Summer visits makes great sense for the student and his or her family as the student may be attending a summer program in the US. Additionally, if a student is already in the US, a summer visit would not require an additional trip to the US, or any time missed from school. The technical answer is yes; schools will welcome students to have an official tour and interview the summer before the student is applying. For the same 8th grader described above, a summer interview will take place during the summer after the student finished 7th grade, but before they entered 8th grade.

Students are usually only given one opportunity for an official interview, and like everything else you have learned, preparation is key. The strategic answer is that having an official visit during the summer is not recommended for most students for several reasons. There is usually significant growth that happens with children the summertime and into the fall months before they apply to boarding school. Many students blossom during the fall of their 8th-grade year. Understanding the need to continue to build their brand, they may take on leadership positions at school, excel in academics and sports as they mature. While application materials will capture this, interviewing before students have had these experiences does not allow the student to communicate this directly to admission officer during the ever so important interview. Let's go back to the superstore analogy to understand this better. If the admission officer is evaluating a product, and he is not aware of specific features that are still in development, he is not able to learn about them and ask questions about how they work. Students who interview too early do not give themselves the best opportunity to promote their brand. Parents, too, will likely see the maturity and growth in their child that occurs during this time and will want to be able to share this growth in their parent interview (more on this later).

Most students, while they may have taken an SSAT in the spring, will not take their first official SSAT until September or October of their application year. While an SSAT is not the only factor in determining the school list, it certainly plays a part. Schools that you may visit in the summer may be out of reach after you are aware of your SSAT, or if a student has done well on the SSAT, they may be able to reach a little higher and apply to more competitive schools provided that the school is a fit.

Students may enroll in a summer program that helps to prepare them for boarding school or enhances their interests and experiences, making their profiles more attractive after the program. Summer visits may not allow students to capture this activity in their interviews.

While official visits in the summer can be convenient, they do not provide the best opportunity for students and families to present themselves from their strongest position. Likewise, during the summer months most schools are shut down, with few, if any, students and faculty on campus. This can make it difficult to get a feel for the atmosphere and community of the schools. For these reasons, we encourage families to interview in the fall or early winter of the application year.

Late to the process? You will need to check each school's policy, but most schools will allow official school visits for a few weeks after their application deadlines.

Inquiry and Pre-Interview Questionnaires

Before requesting an appointment to visit a particular school, parents or students should complete the online inquiry/request information portal found on the admission page of each school that they plan on visiting. The online inquiry usually contains necessary demographic information and may include a few general questions about student interest and activities. Many schools have a second level of **Pre-Visit Requirements** before allowing students to schedule an appointment. This second level of information requested is usually a more detailed survey to gather information about a particular student's interests and activities. These interest surveys can be quite comprehensive. Schools use the information provided to help match a prospective student's interests with current students in assigning the best match for a tour guide on campus. Some schools may also use the information to "guide" the interview. Regardless, preliminary visit surveys can remain part of the application, so care should be taken when completing them.

Once you complete an online inquiry and/or pre-interview questionnaire, schools know much more about you. The information you provide allows schools to target future communications specific to your interests. Students will find that once they have completed these inquiries, they will receive emails from various department heads and coaches in areas where interest was expressed through the questionnaire. For example, if a student expressed an academic interest in History, the History Department Chair might send an email about an exciting speaker that was on campus, about a topic that is currently being taught in a history class or about a current event. If you indicated that you are a basketball player, the basketball coach may reach out to you and provide you with some information about the team. Keep in mind that these emails are all electronically generated based on your responses to the questions on the online inquires and pre-interview questionnaires. Nonetheless, schools can track whether you opened the email. More importantly, these emails represent an opportunity for you to express interest in a particular school and to connect with someone interested and knowledgeable about something that relates to your brand. Students should respond to these general emails and begin to start a dialogue with the individual faculty or staff. In doing so, you are building your brand recognition and, hopefully, creating a loyal customer base to support your brand in the application process. The more people that you have at a particular school who like what you have to offer, the more support you may get in the application process.

Fit is a two-way street. Families, too, want to be evaluating their interactions to see if your brand will thrive and grow in the environment with these particular faculty or programs. Some of the faculty and staff that you interact with may be present and impactful in your life at a specific school, so you want to make sure that they are a fit for you. Finding fit in faculty is particularly important in sports and arts because you may have the same coaches or conductors/ instructors for multiple years. Whether or not you connect with a particular key faculty member and feel that you can work together to accomplish your mutual goals is an important determinant for some students when it comes to fit. Starting to develop relationships with some of the key people in the superstore will become important later when we discuss more specifics about school visits.

Pre-Qualifying Interview Requirements

As you have probably already concluded, many resources go into hosting a student and their family on campus for several hours. From the first inquiry a parent or student makes to the administrator scheduling the interview and tour, to the student and parent interview; all require manpower and resources. Some schools have upwards of 1,500 to 2,000 applicants visiting each admission cycle, which can strain resources in a school's admission office. On the family side, there is also a significant investment of time and resources that come into play when visiting schools. The airfare, travel, hotels, and meals required for this type of visit are costly. Also, some students will need to take some time off from school, making it very difficult to catch up and maintain good grades when they are away. Parents also may need to take time off from work and arrange childcare for younger siblings. All in all, it takes a lot to plan and attend an on-campus interview for all parties involved.

Some schools do not have the staffing to support everyone coming to campus. For these reasons, some schools have a policy where they put in place pre-requirements before students are invited to come to campus for a tour and interview. These requirements most often apply to international students from areas where there is a high volume of applicants, like China. Schools also want to respect the time and effort of families to come to visit a school to avoid families coming to their school if the student will likely not meet minimum requirements. Some schools may employ one or several of the procedures outlined below to help ensure that the students they invite to campus are qualified candidates for consideration.

Pre-Qualifying Video Interviews

Some schools require students to partake in a video interview and screening process with a 3rd party. Depending on the 3rd party, this process usually includes a video interview with the candidate, some assessment of English-speaking ability, and a writing assessment. It is up to the student to schedule this appointment with the 3rd party and to pay for these services. Once completed, the 3rd party sends the materials they prepared to the designated admission office. The admission office then reviews the information and evaluates whether or not the student will be "invited" to campus for a tour and interview.

Pre-Screening TOEFL and SSAT Submissions

In addition to or as a stand-alone requirement, some schools require students to submit a TOEFL or SSAT prior to being invited to interview. Again, the schools are trying to make sure that they invest their time in students who are likely to be successful at their school and not waste the school's or family's resources, inviting students who may not be prepared academically or may not have the English ability to do well. Sometimes this requirement can pose a logistical challenge.

TOEFL tests are available regularly and are probably the most accessible vehicle to comply with this request. A TOEFL test is valid for two years after the test date. Schools have different TOEFL thresholds as a cut-off, but generally speaking, the highly competitive schools may require a TOEFL of 110+ while the slightly less competitive may be looking for students with a TOEFL score of 95+. SSAT is not offered as regularly as TOEFL, and the September SSAT is not usually available outside of the US and Canada. Waiting for an October SSAT to send to schools to meet this requirement can delay scheduling a school visit. Some schools will accept an April or June SSAT from the prior SSAT cycle/year.

Families can be caught off guard by these requirements, which can jeopardize a student being able to meet the deadlines of a particular school. The 3rd party interview companies are based in the larger international cities, and are often booked well in advance. It is helpful to be aware of these requirements early on and to prepare accordingly with early preparations for TOEFL/SSAT and interview. To find out whether a school has these types of conditions, review the school's admission pages under "International Student Applicants" and if they do have pre-qualifying or pre-screening requirements, they will be clearly stated, and procedures and instructions will be provided.

SCHEDULING SCHOOL VISITS

Once families have confirmed a school list, dates of travel, and completed the preliminary visit requirements for each school, they are ready to schedule their school visits. Families should map out a preliminary travel route, and plan to spend 2 hours touring and interviewing at each school. Keep in mind that due to afternoon sports competitions, Wednesdays are usually half days at boarding schools, so only morning visits and interviews are generally available. Some boarding schools that have Saturday classes offer Saturday visits and interviews, but usually only in the morning due to athletic competitions on Saturday afternoons as well. Other than Wednesday and Saturday, most days families can schedule 2 school visits perday, one in the morning and one in the afternoon depending on the location of schools. Boarding schools do not host visits on Sunday. Once you have made appointments to visit and interview at a particular school, it is time to start to market your brand in anticipation of your school visit.

SPECIAL INTEREST VISIT REQUESTS

Determine where your X-Factors have the most impact and get to work to market your brand. If you have developed any relationships through any of the email exchanges that were generated by the pre-interview questionnaires, reconnect with those faculties and/or coaches. Write the faculty or coaches an email to let them know when you are coming for a school visit and request to meet with them. For those that you have not yet been in contact with, you will need to make a "cold call." If you are not familiar with the term, cold calling is a term used in sales and marketing where a salesperson solicits business from potential customers who have not had any prior contact. To prepare for your boarding school cold call, research the school's website faculty page/directory to obtain the contact information for the faculty and coaches who are responsible for programs and teams to which you feel you can contribute. These are your potential customers and loyal supporters. Compose an email describing your X-Factor, include any support about how well your X-Factor performs (multi-media links, portfolios, sports CV's). Indicate how your X-Factor can add to the school community. Request to meet with this person during your school visit to pitch your brand. As much as you must market yourself, it is also essential that you meet with the faculty in the areas that are important to you to learn about how that school's particular program fits you. Remember: Fit should always be a two-way street.

2.3 STUDENT AND PARENT SCHOOL VISIT

Students can learn a great deal about the facts and offerings of different boarding schools by spending time on the school's website and talking with current and past students. However, there is no substitute for being on a boarding school campus and being able to "experience" a school first-hand.

SCHOOL VISITS

Finally, it is time to see what all the fuss is all about. Planning for school visits is exciting. With a little advance planning you can make your visit the most successful it can be. It is time to decide everything from what you will wear on your school visits to who else can you reach out to at the school to learn about you and your brand.

What to wear?

Each school has its own rules, otherwise known as dress code, for how students should dress in class and out of class. Some schools are quite casual and allow students to wear blue jeans and t-shirts. Other schools require boys to wear a jacket and tie and girls to wear what would be considered business casual style dress. For some students, the dress code requirements of

a particular school become a factor of fit. Some students prefer schools that have a more relaxed dress code, and others feel more comfortable in more formal attire. The fit of the dress code only comes in to play AFTER you are admitted. On the application side of the process, students should plan to dress in the following way for their school visit, even if a school has a relaxed dress code.

- ***Boys:*** *Sport or suit coat/blazer, tie, slacks, and comfortable dress shoes. No jeans, t-shirts, or open collar shirts.*
- ***Girls:*** *Nice slacks, dress, or skirt (not too short) with a shirt or blouse. If you are wearing slacks, plan to wear a blazer over your shirt as this is usually a dress code requirement. Do not wear any thin strap shirts.*

Campus tours last around 45 minutes and will involve a lot of walking, sometimes on uneven terrain. Wear comfortable shoes, but do not wear sneakers. The weather can be quite unpredictable, and you should pack a raincoat or warm coat depending on the time of year you are visiting. If you are touring east coast schools during the late fall, early winter, you may need a pair of boots, a warm hat, and gloves. Some students may wish to wear their school uniform blazer to school visits if they have one.

Oral hygiene is essential in American culture. Make sure to brush your teeth before your visit. If you have eaten a meal since you brushed your teeth, you may want to plan time for a quick visit to the restroom in the admission office before your tour/interview to make sure that your teeth are still clean. Perhaps pack a portable toothbrush!

What to do?

Admission Arrival

Student ownership is encouraged every step of the way. Arriving at the admission office for your interview is a perfect opportunity for students to demonstrate ownership of the decision to apply to boarding school. All

admission offices have a reception/check-in process. There is usually a desk or office at the entrance to the admission office staffed with a friendly admission person or sometimes a student volunteer. Students, not parents, should announce themselves to admission reception. Students should clearly state their name and the time of the appointment. Surely, the admission receptionist will welcome the student, and it is at this time that the student can introduce who is accompanying him/her on the visit — your parent(s) and siblings (if any are along for the trip). This initial interaction is also a good time for the student to inform the admission office of any additional special interest meetings they have secured with faculty/coaches through building their relationships or cold calling. Keep in mind that most faculty wear multiple hats at boarding schools. While they may be the varsity basketball coach, they are also likely teaching a class and may not be available at the time you are visiting campus. Having said this, it doesn't hurt to alert the admission reception to any unsuccessful attempts to schedule with certain faculty. The admission receptionist may able to assist in facilitating an impromptu in-person meeting while you are on campus. Some admission offices may be very willing to help you in this manner, and others may not.

Admission Waiting Room

After checking in for the appointment, most families will be asked to wait in the admission reception area while tour guides are gathered and assigned. If you are on time, this is usually no more than 5-10 minutes. Use this time to visit the restroom and check on any untucked shirts, etc.

Most admission offices have at least beverages to offer families, and some have light snacks. Drinks usually consist of water/juice and some high-tech hot beverage dispenser machine. Resist the urge to make yourself a hot chocolate;

water is the safest bet to keep you and your clothes clean throughout the visit. If it is cold outside, have hot water while you wait to go on your tour! Save the hot chocolate for your way out, after your visit.

Tour

You have invested the time to include a particular school on your shortlist and to visit. Before arriving at a school, you should have done your research and know as much as you can about each school. Your research has resulted in an understanding of the programs and activities that interest you. You know what types of classes are offered and have explored the electives you can take. What is important to you in determining fit? You should give some thought to this issue before you arrive for your tour and interview. Why do you want your brand to be represented at this particular school? How will you best fit in, and where can you explore and grow to meet your goals? Compile a list of questions that are important to you in determining if the school is a fit for you. For some, it might be academic rigor; for others, it might be proximity to a major metropolitan area so that they can enjoy cultural events. For others, it is essential that they learn about the school culture and what students do on the weekends. Others may want to know about dorm life and what types of living arrangements students can choose. The student tour guide is a perfect resource for you to gather the information and form some answers.

Most schools require students to give back to the school community in one way or another. The role of tour guide is usually one of the choices that students have for their "job." However, student tour guides are often a select group of students. They have undergone a screening and training process to provide this service to you. Keep in mind that they are very proud of their school and that they are likely giving up their free period to show you around their campus. It is also worth noting that some admission offices have a procedure for student tour guides to provide some feedback to admissions about their interactions with you are on tour. The student-lead tour is a marketing opportunity for you! Seize it!

Having information about how a particular school fits you will help you to ask your student tour guide the most meaningful questions. Students should compile a list of 10 or so meaningful questions and be prepared to ask them of the student tour guide at each school they visit. The emphasis is on meaningful. Anything that you can find easily on the school's website would NOT qualify as "meaningful." Don't waste the opportunity to get a student's perspective by asking questions like, "How many students are there in an average class?" (Unless, of course, that information cannot be found on the school's website!)

Students and parents may go on a tour together. Students should interact with the student tour guide on tour. Nothing is worse for a student tour guide than showing someone around who doesn't seem interested in their school. When a tour guide shares some information, you should enthusiastically respond, showing them that you are engaged in the tour and like their school. Find out about the tour guide and what they like about the school. Share with them some of your interests and, hopefully, you will find some common ground to continue your conversation. Discuss the school tour with your parents beforehand. Parents should not monopolize the student tour guide with questions. Perhaps you have a "safe" word between you and your parents. Use this word during the tour to remind your parents to hang back and let you interact with the student tour guide to establish a relationship. An occasional question from Mom and Dad is fine, it shows interest and parents need to be informed as well. However, when parents have the majority of interaction with the tour guide, it prevents the students from interacting and conveys that the ownership of the process may not be driven as much by the student as it is by their parents. Parents will have plenty of time to ask questions of the admission officer later on during the parent interview.

Tours usually last about 45 minutes to an hour and involve a pre-determined route around campus under a time constraint. The tour is intended to showcase the highlights of the school. Not all buildings and programs can be on tour. If you have a particular interest, share this with your tour guide at the beginning of the tour. Chances are your interest area will be on the planned route, if not, an excellent tour guide can try to customize the tour to show you that area of interest. If time does not permit, the tour guide can provide you information about how to visit this area on your own after the official visit. For example, "I am so excited about the tour. I am interested in robotics. Are we going to be able to visit the robotics classroom?" Remember, fit is a two-way street. You have come a long distance to visit the school. Learning about and directly experiencing programs in your areas of interest is an essential determinant of fit.

What to Say?

Interview

The interview is a culmination of the preparation to apply to boarding school. Students who start the planning process early should have a lot of information about their brand to share with the boarding school admissions representative during an interview. Before interviewing, a student should know their brand, what it represents, and how to communicate it to a boarding school. The interview is one of the vehicles and, arguably, the most effective one to demonstrate this. The interview, or more accurately the conversation that takes place between a student and an admission officer, can make or break an application. Time and time again, seemingly qualified students whose interviews are substandard can be denied admission. Other students who are perhaps not as strong candidates academically, yet had a fantastic interview, often can be offered admission over a stronger candidate.

Since most application deadlines are early January, most students will not have completed their applications before interviewing. However, their profiles and stories should be well established and integrated. The interview represents the student's opportunity to tell their story, show their personality, and sell the admission officer on the fit of their brand.

Formal interviews are usually reserved for students who are applying in the coming year. Interviews can last anywhere from 20 minutes to 45 minutes. Duration seems to be a factor of two variables. The first variable is how much time an admission officer has to devote to the interviewee. During the busy times when many students are visiting, back to back interviews occur, preventing admission officials from being able to spend extra time. Generally, if an admission office invests extra time to interview a student, it is a good sign that the student has attractive information to share. If the admission officer spends more time, they are interested in learning as much as they can about your brand. You are the key! Effectively selling your brand during the interview generally means more interest from the admission officer!

Many students and parents want to know precisely what questions are asked in the admissions interview and what are the "correct" answers. A good interviewer does not have any set questions, and a good interviewee can provide spontaneous responses to any question an interviewer may ask. Some interviewers may ask a seemingly random question like, "If you were a color, what color would you be?" or "If you were a molecule, what type of molecule would you be?" Boarding schools are a place of conversation and interaction, and specific responses should not be scripted or practiced. A successful boarding school student is one who can think on his or her feet, adjust his/her approach in a conversation and respond appropriately. A skilled interviewer can pick up on an over-rehearsed student and most likely will not be impressed by the candidate.

That being said, there are some general themes that most interviewers seem to cover. Learning about how a particular student deals with challenges, what types of interests and passions does the student have, and why they are a fit for boarding school are just a few. Interviewers will encourage students to talk about their particular interests and passions. They are interested in a student's learning style and how they see their learning style as a fit at their specific school. Students should think about their responses to these types of questions before going on school visits.

Most interviewers will start their interview with some ice breaker asking the student a benign question like, "How was your tour?" This is an opportunity for students to complement the admission officer about their school. Remember the analogy of a boarding school being a family? That comes into play here again. Boarding school faculty and staff are very proud of what they have to offer, like you may be proud of your home or your room. When you have guests to your home, you are likely to appreciate a compliment. The folks who live and work at boarding schools are no different; they are very proud of the programs and facilities of the boarding school that they call home. Be sure to be complimentary.

All boarding schools have something to offer students. While on your tour, look for something in particular that you were impressed with at the school.

Perhaps the particular school has a fantastic swimming pool, or you observed a specific lesson in a classroom that you found exciting or a friendly interaction between students caught your eye. Find something meaningful so that you can comment specifically as a response to the ice breaker question, about your tour.

To effectively interview, students need to understand the interview mentality that best compliments the application process. Interviewing is a little bit like the 19th-century British fairy tale, "Goldilocks and the Three Bears." In this story, a naughty girl named Goldilocks happens upon the home of a family of bears. Papa Bear, Mama Bear, and Baby Bear have gone for a walk in the woods while their morning porridge is cooling. A hungry Goldilocks boldly enters their home and tries each one of their bowls of porridge until she finds exactly the right bowl, the one that is not too cold, not too hot, but is just the right temperature. Boarding school interviews are a little bit like this story.

Too Cold

An interview that is "Too Cold" would be one where the student does not engage in the conversation and provides very little information to the admission officer. Here, the admission officer finds themselves driving the conversation trying to coax the student into talking about themselves. Even the most talented student on paper needs to be able to present themselves during the interview process. A "Too Cold" interview is usually the result of a situation where it is either the student's first interview, the student is very nervous, or the student has not prepared well.

Too Hot

While an interview with a student who does not readily provide much information is not considered an effective interview, a conversation with a student who offers too much information in a rapid-fire manner is also not an effective interview. In these instances, students who provide too much information can be off-putting and come across as ingenuine and overly scripted. They are "Too Hot." Most interviewers will try to encourage this type of interviewee to slow down by asking a question or trying to change the subject, hoping that the student can pick up on some social cues. Most kids don't recognize this, keep on their agenda, and eventually burn out, providing the interviewer with all of their information within a span of 5 to 10 minutes. A "Too Hot" interview leaves the interviewer with the impression that the student is not able to pick up social cues. This type of monologue delivery also gives the idea again that the student has over-prepared and rehearsed. This type of interview can also give the impression that the student may not be the type of student who would readily engage in collaborative dialogue and discussion. A "Too Hot" interview is a one-way delivery of information.

Just Right

"Just Right" is a balance of the student speaking and the admission officer speaking. It's a true dialogue, with each asking and answering questions. While this type of student has undoubtedly prepared for their interview, they understand and can participate in an interaction consisting of a give-and-take conversation. Unlike the "Too Cold" interview, where the interviewer is trying to guide the conversation and the "Too Hot" interview, where the student is the only one delivering information, the "Just Right" interview, is a two-way street. The student provides some information to their interviewer, allows for questions, and provides impromptu responses. In the "Just Right" situation, the interviewer feels the pleasant interaction of being engaged in a conversation with an interesting student who can talk about and compellingly promote themselves.

Students who are "Too Cold" give the impression that they will not be able to add much value to the classroom because they don't contribute and interact. Students who are "Too Hot" can initially enjoy status in the classroom by being able to deliver information, but eventually either run out of information or other students become frustrated by their monopolization of the conversation. "Too Hot" students are not seen as team players, which is an essential skill of a boarding school education. The "Just Right" interviewee recognizes when it is time to provide information, such as when an admission officer prompts a question, and when it is time to listen and allow the admission officer to speak. It is the give-and-take that is essential to excellent communication in general. Good communication skills, in particular, are crucial to a boarding school classroom where the foundation of teaching is based on the Harkness Model of discussion-based learning.

Some Tips for Student Interview

The Handshake

Americans love strong handshakes. Like everything else, the handshake is something to be practiced. A little bit like the Goldilocks analogy. Practice a handshake that is firm but not overwhelming. A weak or floppy handshake should be avoided at all costs.

Eye Contact

When you greet the interviewer, look them in the eye when you shake his/her hand. Also, look the interviewer in the eye when engaging in conversations during the interview. Eye contact conveys that you are confident in what you are saying and are engaged in the conversation.

Avoid "yes/no" responses

The interview should give the school a sense of who you are. Practice thoughtful responses that reveal something about yourself rather than a monosyllabic response.

Do Not Memorize Your Responses

Think about what to say about some fairly typical questions and practice having a conversation with someone about them, not rehearsing a statement. The best interviews are conversations that flow in a give-and-take manner rather than a stiff practiced response. If you have prepared well, you will naturally know the answers to all of these questions and ones yet to be asked because you know your brand and what it represents.

Practice

Ask an adult to spend 30 minutes asking typical interview questions like the ones below, and then offer you constructive advice. Incorporate that advice and then find another adult to practice with. Relatives and teachers are excellent sources for this practice. Some students are very comfortable talking about themselves with adults; others benefit from many practice sessions.

Some typical types of questions

- ***"Tell me about yourself."*** What makes your brand special? Highlight talents, interests, experiences, and goals. Do not provide too many details to this question. Provide an overview and allow the interviewer to respond with a follow-up of more specific questions about the information you shared.
- ***"Why do you think that US boarding school is a good place for you to continue your studies?"*** You need to have given thought to your response here. Think about how you may contribute and the

growth you are anticipating as a student at boarding school. Who will you be as you approach graduation? What types of challenges and development are you looking for in this experience?

- ***"Why do you want to come to XYZ school?"*** Focus on something specific about the school and the offerings. For example, "I enjoy playing the cello, and I learned that your orchestra is one of the best. I am hoping that I can try out." Make sure to do your school-specific research to know your customers.
- ***"What subjects are you studying now?"*** Give positive answers to any question about your current school, subjects, or teachers. Don't criticize or compare your current school to the boarding school. Try to talk about any specific academic interests or give the interviewer an understanding of your school curriculum if they are not already familiar with your school.
- ***"Do you have any particular academic interests?"*** Here is your opportunity to discuss your academic X-Factors. Give a general overview of your academic X-Factors to set the stage for a conversation with the interviewer, rather than to describe each one in detail. The interviewer will be sure to follow up with a question that will allow you to further elaborate.
- ***"Tell me a little bit about the activities that you are involved in at school."*** Here is another opportunity for you to talk about your X-Factors as they relate to school activities. Here is where your research comes in particularly handy. Know what types of programs and clubs are offered by the particular school that relates to any current activities you wish to continue or even try at boarding school. If your current school does not provide these activities, express how you plan to pursue these activities at boarding school. Also, if you have emerging interests, convey your interests, and find areas where you can potentially contribute.

- ***"What do you like to do in your free time or on the weekends with your family and friends?"*** Here are those X-Factors again — you can see why they are important.
- ***"Share with me a situation where you had to work hard to achieve your goal?"*** Here the school is trying to get a sense of your resiliency to overcome challenges. Think about a meaningful challenge you had, such as trying out for the school play or preparing for a debate competition. It is best here to avoid academic examples such as when you failed a math exam or did not get an A on a project. Students who convey that they are overly concerned about academics may raise a red flag that they are the type of student who may not readily involve themselves in programs and experiences outside of the classroom.
- ***Scenarios. "Tell me about a time when you had to make a difficult choice between two equally good or two similarly tricky situations. How did you make your decision?"***
- ***"Do you have any questions for me?"*** This is another opportunity. Seize it! Make sure that you have something to ask. You should prepare a list of 5-7 questions for each school. Asking questions demonstrates that you have given thought to and are invested in the process of finding the school that best fits you.

To Gift or Not?

It is discouraged for students to bring gifts for admission offices and student tour guides.

Post Interview

Students should send a thank you note to their interviewer and tour guide at each school they interviewed. Avoid a generic thank you. Every communication that you have with a school can be logged in to the school's application management system and is representative of your brand. Students should take care in composing every communication as it is reflective of you as a student and a person. For the thank you note, add some specific details that the interviewer or student shared with you. Make reference to something specific about the school, and how it impacted or impressed you. Use the school name in your correspondence. Be careful to proofread all communications for grammar and MAKE SURE that you do not confuse school names and details. There is NOTHING WORSE than when a student sends correspondence to one school referencing another. Parents are encouraged to have a look at the communications that their children send to try to avoid these costly mistakes.

PARENT SCHOOL VISIT

The importance of the student school visit and interview was discussed in the previous section. One would naturally expect students to be front and center in the admission process to boarding school, and they are. However, parents should be aware that they are also considered part of the boarding school application process!

At most schools, parents are expected to participate in the tour and will also have the opportunity to speak with admission staff. Boarding schools interview parents first and foremost because they want to learn more about the child who is applying. Parents can provide a perspective to the application that is hard to communicate in the application. Each school is interested in knowing that you, as a parent, understands their school and that your child wants what the school has to offer. They also want to know how serious you are about their school, versus the other options/schools that you and your child are pursuing. Schools are also trying to get a feel for parent fit. They are trying to gauge how easy or difficult you'll be to work with if issues arise later in your child's school career. They are also assessing whether or not your "entire" family will be a good fit for their community. While the admission interviewer is evaluating whether your child and family are the right for them, parents should be observing, asking questions, and making the same determination on their end. "Is this school a good fit for my child and family?"

What to Expect?

One of the most stressful components of the process for parents is the responsibility they feel to find the school that best fits their family's values, philosophies, and expectations. The school visit is by far the best way for parents to determine whether a particular school is a good fit for their child and family.

Many parents ask if both parents are required to attend the school visit and hence participate in the parent interview. Naturally, there are families where only one parent can attend the visit due to work commitments, perhaps caring for younger siblings or aging parents at home or the situation where the parents are divorced. A student's application will not be adversely affected if only one parent attends the school visit and parent interview. However, it is probably no surprise that if possible, the best scenario is for both parents to attend. Both parents in attendance helps to convey how important this decision is for the family. Also, when both parents visit, they share the experience and can discuss the particulars of a school as a fit for their son or daughter. When only one parent visits, the visiting parent has the responsibility of communicating their experiences and impressions to their partner, perhaps giving greater responsibility on school decision to the visiting parent.

Some parents of international students may not be entirely proficient with English to be comfortable participating in the parent interview. If one of the parents is more comfortable in English than the other, this parent would be the preferred parent to visit if both cannot attend. Some families share concerns because neither parent can participate in an interview in English. Some schools can accommodate a language barrier by providing an admission counselor that does speak their language, providing a translation through a teacher or by inviting the student into the parent interview to translate for his/her parents.

The term Parent Interview is a misnomer. The more accurate way to think of it is as a parent meeting. Having sat in on hundreds of parent interviews while accompanying students on school visits, a well-seasoned admission officer at a top 10 school described the purpose of the parent interview perfectly. She described the parent interview as “an opportunity for parents to tell her something about their child that she could not read in the application.” This simple statement should frame every parent’s meeting with an admission officer. Parents should use the parent interview to help the admission officers to understand their child better and to find out more about the school.

Most parent interviews occur directly following the student’s interview. Most are meetings between the same admission interviewer that interviewed the student. However, a few schools have a family meeting where the parents join the student interview at the end. In this situation, parents and students meet with the admission officer together. Typically, parent meetings will last about 15-20 minutes.

Please try to make arrangements to leave younger siblings at home. The rigor of the schedule required for school visits is very demanding, and young children find it difficult. There is nothing of interest for younger children, and they serve as a distraction — all attention should be on your child who is applying.

What to Wear?

Parents are recommended to dress comfortable but adhere to business casual attire. A blazer is suggested for fathers. While a necktie is not necessary at most schools, it is considered appropriate if desired. Since most tours involve a campus walk, comfortable footwear is important as well as clothing to accommodate cold or inclement weather. Avoid attire that is too flashy or pretentious.

What to Do?

Research

All boarding schools are different. Like students preparing for their interviews, parents should do their homework too. Parents should formulate a clear idea and be prepared to articulate why the school made your child's list and what makes it an excellent fit for your child and family. Before each school visit, it is encouraged that parents do their research, and parents should know at least the following about each school:

- ***Mission and philosophy of each school***
- ***Programs and extra-curricular offerings that are important to your child***
- ***University matriculation — so you don't ask in the parent interview!***

Brainstorm Questions

Remember the purpose of parent interviews: The admission officer will ask parents if they have any questions. Parents should brainstorm questions to ask the admission officer. Also, be prepared to answer a few questions, such as: Tell me about your child or how did you learn about our school?

The exercise of brainstorming and writing a parent statement can be beneficial in preparing for the parent interview. Parents should at least have worked on a draft of their parent statements before going off on school visits. The discussions that occur in preparing a parent statement can be quite similar to those that can happen in a parent interview. Writing a draft parent statement before visits gives parents time to process and formulate their responses. Do consider some of the challenges that your son or daughter has had. Do understand that no child is perfect and that being able to identify areas of growth for your child shows that you are parents who understand that boarding school is a vehicle for such growth.

Putting together a draft of your parent statement before your parent interview also helps to provide a consistent message between parent statements and interviews. If possible, both parents should have input into the parent statement, so parents are aligned during the parent interview. Keep in mind that while parents want to put their best foot forward to present their child as attractive as possible to admissions, the parent interview is also intended for parents to be able to determine if a school is a good fit for their child. The best way to find that out is by visiting a school and asking questions.

Interview: Get Prepared Physically and Emotionally

Make sure that your child is healthy, well-rested, and fed before the interview. Staying healthy is easier said than done for most international families who have traveled a great distance to attend school visits. If possible, try to arrive in the US a few days before the first school visit to acclimate to the time zone, climate, and recover from jet lag. Also, if possible, schedule high priority schools in the middle of the visit week. This way, your child will hopefully be feeling less jet-lagged and a bit less nervous and more practiced. Having a supply of familiar snacks can help a student to get through if they don't have time for a meal.

Pace yourself so that you arrive at school close to the interview time. You don't want to sit around in the admission waiting room too long with a nervous child, and you also don't want to rush. Planning to arrive at a school 15-20 minutes before the appointment time should give you enough time to locate the admission office and get settled.

While in the admission office, engage in friendly conversations with other parents and students, but avoid the "comparison game." Prospective parents may ask other parents or students about SSAT scores or what other schools they are considering. Stay away from this type of exchange! This process is nerve-racking enough without worrying about how your child stacks up against his neighbor in the admission waiting room or the ones you saw on tour.

Make sure you're helping — not hurting — your child's chances.

Tour: Pay Attention

All schools offer parents a tour as part of the school visit. Most schools have parents and students tour together with a student tour guide. Some schools will have families tour with other families in a group. Other schools pair parents with a student tour guide and provide tours for parents only. Still, other schools may have a current parent tour a prospective parent. Even further, some schools may provide tours in different languages for international parents.

If you find yourself invited on tour with your child, think of yourself as a guest on tour. This tour is intended for your child. If your child has prepared well, he or she knows that the tour is an opportunity for him or her to not only learn more about the school but to also get to know a current student. It is important to remind everyone again that some admission offices do seek feedback from their student tour guides. Student tour guides are coming from the perspective of reflecting on whether or not they would see a student as a fit at their school. They are evaluating based on questions like: "Would I want this prospective student in one of my classes or on my team?"; "Do they seem like they would be a good dormmate or roommate?"; "Are they interesting, and would make a good dinner companion at sit down dinner?" Students should prepare questions and interact with the student tour guides. Parents should resist the urge to take over the tour and ask the majority of the questions of the tour guide. Parents ought to hang back, within earshot, and walk behind your child and tour guide. Allow your child to take responsibility for the process and the privacy to interact and show the tour guide what type of peer they would be. Pay attention but save your questions for your parent meeting with the admission officer.

School Culture

Be observant on tour for a few reasons. First, you are determining if a school is a good fit for your child. In addition to seeing the facilities, pay attention to the interactions between students, with yourself and other students. Do you see faculty on your tour interacting with students and other faculty? How do those interactions seem? Do students welcome you by making eye contact or saying "hello?" Being observant will help you to get a feel for the culture of a school — something that viewing a website will not afford you.

Dorms and Food

Your child is going to be spending a lot of time in the dorm. Be sure to have a look at the dorm rooms and peak in the bathroom of the dorm. Food is important. Most school tours include a tour of the dining hall. Make a mental note of the types and variety of foods that you see. Ask your tour guide about their opinion of the food or maybe even about their favorite meal. If your

child has allergies, ask the admission officer to describe the protocols for accommodating allergies or have the tour guide show you where the gluten-free breads are kept as well as the dairy-free products, for example.

Supervision

It can be even more important to international families that boarding schools take their role as "surrogate parents" very seriously. Ask about the Code of Conduct and how supervision is implemented on a 24/7 basis. Different schools have different policies about discipline, especially when it comes to major offenses like academic dishonesty or drugs and alcohol usage. Depending on the school's location, schools also have different policies about leaving campus during the day and on weekends. Find out about the rules for leaving campus to go to the mall with a friend or home with a day student. Boarding schools consider your child's safety as their primary concern and require parent permission to leave campus. Even an accidental violation of this rule by your child could result in severe disciplinary consequences. Use this information to discuss with your child and make sure that they understand the rules as they pertain to a fit school for them.

International Student Ratio

Most parents are concerned about the diversity of the international student body. As parents, you are sending your child a great distance to experience a US boarding school education. A school that has too many students from the same region may not be attractive to some parents. Similarly, a school with too few students from their given area may be equally less desirable. Many students with the same background as your child may create an environment where he or she does not meet students from other regions. However, without a critical mass, parents may worry that their son or daughter may not have the opportunity to culturally bond with students from similar experiences. International Student Ratio can be a delicate question to pursue with the admission officer because parents want to convey that they value inclusiveness and diversity in a school community. To open the door to this conversation, without asking directly, you can make a statement about diversity that you observed on your school tour and hope that this opens the door for the admission officer to describe the school's diversity. For example, "I saw that when we entered the dining hall, there seemed to be flags from many different countries represented." Most likely, this statement will elicit the information you are looking for as the admission officer will share further details about the diversity of the students attending. They may even go further by explaining the diversity plan and also providing specific numbers of students by country. Being observant on tour allows you to be better informed to make comments during your parent interview and ask good questions.

What to Say?

Ask Questions

The parent interview is a part of the school visit and helps parents to find out more about the school. The following list of questions will help you dig a little deeper and pick up the important differences from one school to another. The information obtained will help inform parents about how a particular school fits their child. Asking the same questions at each school will help you to compare across schools. Like your child, parents should avoid asking obvious questions that you can learn from the website or admission brochures.

Some questions that parents may want to ask an admission officer. Suggest that you prepare 5-6 that are of most importance to you.

1. ***How is the mission or philosophy of the school carried out in school on a day-to-day basis? (remember to research each school's mission or philosophy)***
2. ***How does XYZ school encourage a student's strengths and help support a student's weaknesses?***
3. ***Do students have the opportunity to try new things? Can you give me an example?***
4. ***How much homework is there on a nightly basis, and what is the general philosophy about homework? Where can students do their homework?***
5. ***How much downtime do students have in a typical day?***
6. ***Describe the typical graduate of XYZ school.***
7. ***How many students leave before graduating, and what are the most common reasons why they leave early?***
8. ***What are your faculty and staff turnover from year-to-year, and what are the most common reasons faculty and staff leave early?***
9. ***How engaged is your parent organization, and what kind of activities do parents contribute to campus life? How does the school communicate with parents and international parents in particular?***

10. What activities/programs do you have to help transition new students into the school?

Schools like the parents of international students to be involved in their school while the child is a student and hopefully afterwards. It is important to ask questions that demonstrate that you intend to be an involved parent.

1. ***How often does the school have opportunities for the parents to visit their children?***
2. ***What type of communication is there between the student's advisor and the parent? What mechanisms are in place if my child is not doing well academically or gets sick? Who will contact me?***
3. ***How can parents get involved in the school community locally and in my home city? (It also never hurts to let the school know that you are willing to help when asked; many schools rely on volunteers, and involved parents are highly desirable.)***
4. ***Does the school provide transportation to and from the airport for school vacations and holidays?***
5. ***What is the policy for the students staying at the school for the shorter breaks like Thanksgiving or Parent's Weekends?***
6. ***Do students need to find a place to stay because the dorms are closed, or does the school continue to provide housing for those students who live far away?***
7. ***Most parents are concerned about health and safety. Ask how these areas are addressed.***

Fill in the Gaps

Most interviewers will begin the parent interview by summarizing information that was provided by the student. During the parent interview, admission officers try their best to set the parents at ease and are usually highly complementary about their child at this stage of the process. Many parents leave their parent meetings feeling very optimistic about their child's "chances" of being admitted to that particular school.

Most parents naturally want to promote their child, and there is a natural urge to boast during the parent interview about their child to the admission officer. Parents are discouraged from listing their child's accomplishments or resume. Rather, listen carefully when the admission officer is summarizing what they learned about your child. 99% of students either forget something important, or they run out of time and are not able to share something important about themselves. If you realize that your son or daughter has not shared something significant about their profile, asking a question is a good way to bring it up. For example, if your child has writing as one of their X-Factors and did not seem to share any information about this interest with the admission officer, you could say something like, "I noticed that you have a student run magazine, did Jimmy tell you about his interest in creative writing?"

Admission officers are busy, and they have a limited amount of time, so if you are going to add any information, make sure it is important and relevant to your child's application process. If your son or daughter forgot to mention that they play the cello for example, that is important. If they forgot to mention they were part of the math team or were cast as the lead in a play, that is important. If they forgot to mention that they received honorable mention in a writing contest in 2nd grade, this is not important. Rest assured that if your child has done a thorough job in their application, most of the essential parts of their profile should be in the application materials when they are submitted.

Parents who insist on running through all of their child's accomplishments and achievements in the parent interview are much like the "Too Hot" interviewee. An admission officer is much more interested in hearing from parents about the child's personality traits that motivate and nourish their child rather than his or her achievements.

Answer Questions

Some typical questions parents may be asked during the parent interview:

1. ***Tell me about your child.***
2. ***Why are you interested in our school?***
3. ***What is your child most proud of?***
4. ***Describe something your son/daughter worked very hard at accomplishing?***
5. ***What does your family like to do together on weekends and holidays?***
6. ***What is your child's attitude towards school and learning?***
7. ***What is your parenting philosophy?***
8. ***Do you have any concerns about your child's readiness for boarding school?***
9. ***What other schools are you applying to?***

Let's spend a few moments specifically discussing this last question #9. Parents should consider your answers carefully if you are asked about what other schools your child is applying to. While schools do track what other schools students are applying to and admitted to, one of the main purposes of this question is to get at whether you know what you want in a school. A school list that seems to have no common thread other than ranking gives the admission officer the message that your consideration for fit is paramount and that your only concern relates to ranking. They are also trying to gauge how their school fits within the overall list. If you are applying to many schools, what is their chance of your child attending their school? As suggested previously, students should approach this process with a balanced and thoughtful list. If they have done so, the school list should include some highly competitive schools, competitive schools along with some less competitive schools. The list should consist of schools that have special programs that fit your child and be of a similar size.

As parents, we are all concerned about making sure, "we don't hurt our child's chances of being admitted to their dream school in any way." All parents will ask if it is possible for a parent to negatively impact the child's chances of being admitted to a school. To that the answer is, "not often, but it is possible, and it has happened." Admission officers can be turned off by the occasional parent. This can play a factor, particularly at either highly competitive schools where there are many qualified candidates with parents who are more aligned or in very small schools where parent involvement may be more likely. Generally speaking, the "bulldozer" parent has the most potential to damage their child's chances of being accepted at a particular school. Boarding schools look at the importance of educating and raising a child as a partnership between the boarding school and the family. The bulldozer parent does not exhibit the partnership characteristics that boarding schools are looking for. Rather, they present themselves as demanding and difficult. They may ask aggressive questions about college matriculation, boast about their child's abilities, and convey that a school must meet his/her needs. A 'bulldozer' parent may also provide so much detail about a student's accomplishments that they cause the admission officers to question about how involved the parent as been in their child's achievements. A parent who gives the impression that have had a strong hand in their child's process can cause doubt on the v of the entire application.

There are many qualified students for each spot in a boarding school, so make sure that you are not the type of parent that encourages the admission officer to look beyond your child for the other child whose parents are engaged and cooperative about the process.

Things to Avoid

1. ***Calling the admission office too often.*** *This usually occurs most often after application materials have been submitted, and parents are checking on the receipt of materials. Always check the school's website or the application method (SAO or Gateway) first. With the online submission of documents, it seems logical that submissions would be available in real-time and show up immediately at a school. However, this is not the case; most schools need two weeks to process materials, so make sure to give them time. Remember the Bulldozer parent!*
2. ***Canceling appointments, especially more than once without a very good reason.***
3. ***Being late for an admission appointment.*** *During busy times of the year, admission officers conduct back to back interviews, so throwing off their schedule might jeopardize their ability to spend the time to interview the student. Additionally, school tours are led by students who are using their free class block to give a tour. If you are late, the student tour guide will also not have the time to provide you with a proper tour. If you are going to be late for an unavoidable reason, call the admission office to advise them as soon as you realize it. If the office allows you to visit still despite arriving late, be sure to apologize when you do arrive.*
4. ***Badmouthing your child's current school and his or her current teachers.***
5. ***Withholding information about the child's needs.*** *If your child needs additional academic support, it is best to identify that upfront to make*

sure that a school can accommodate them.

6. ***Putting excessive pressure on the child to perform or be overly involved in describing a child's achievement.*** *If a parent seems overly invested or knowledgeable about the details of achievement, the admission officer may wonder whose achievement it is after all.*

7. ***Ranking schools.*** *The admissions staff knows that you are looking at several schools and encourages you to do so to determine the best fit. Regardless of whether their school is on the top of your list or the bottom, be cordial and complimentary.*

8. ***Answering questions directed at the child.*** *When parents take over the process for the children, it raises red flags for admissions about the student's investment and ownership of the process as well as their ability to self-advocate. Let your child speak for themselves.*

9. ***Trying to impress the staff with your business or social contacts. Schools value diversity and look for students from many different backgrounds.*** *The application allows an area for parents to indicate your business. If you do have family or friends who have attended the school, or have a connection to a school, you can casually mention this during the interview, and there is also a section of the application to indicate this. The best approach is to have the connected individual contact admissions on your behalf and introduce you before you visit.*

10. ***Asking how many students are admitted to Harvard or any particular prestigious school.*** *The college matriculation list is available on the school's website. If you must ask about the college counseling process at the school, the following tone is suggested:*
    - *What year does the college counseling process start?*
    - *What distinguishes your school in the college counseling process?*

11. ***Asking what your child's chances are of being admitted to a particular school.***

12. ***Saying anything too negative about your child.*** *The counselor might ask if you have any concerns about the readiness of your child to attend boarding school or to share a challenge that your child has faced. No child is perfect, but it helps to prepare your response to these questions ahead of time. Identifying some areas that your child may have faced some challenges and provide antidotes on how he or she has been making strides to improve in this area is a good strategy if asked this question (This information is usually helpful for your parent statement, so preparing a draft of your parent statement as recommended will help you to respond to this question).*

13. ***Telling each school that it is your first choice so as to make it seem like you're more invested than you might be.*** *It is perfectly acceptable to inform them that you're looking at and comparing a few other schools as choosing the right school is an important decision. If after you have visited all schools on your list, you know that a school is truly your child's first choice, then it is prudent to let the admissions committee members know. Students should be genuine in their comments and reserve this endorsement for a note to the admission officer only after you have visited all schools. Communicating school preferences will be discussed later. Remember the importance of ownership. School preference is best communicated by the student.*

14. ***Being overly enthusiastic.*** *At this stage, most feedback is very positive, sometimes effusively so that parents leave the parent interview feeling overly confident. It may be obvious that admissions like you and your child, but don't get carried away or be overconfident. A smile, a polite handshake, and thank you are all that is necessary after the interview.*

Remember

Students should take the business card (name card) of the admission interviewer. Students will need to write the interviewer a thank you note within a week of their visit. It is a good idea for parents to ask for the name card of the admission officer, as well. Chances are, your child may misplace the card given to them by the admission officer, and then you are left searching the website to find the admission counselor who conducted the interview. It is also perfectly fine to provide the admission officer with your name card after the parent meeting.

Think about visiting a boarding school like you would visit someone's home. Be cordial to everyone you meet, from the time you enter the school until you leave. Don't bring samples of your child's work to share with the admission officer — your child should include these in his/her application. Make sure to turn off your cell phone and be present in the process. Finally, do what you can to make this process as easy on your child as possible. The more prepared you are, the more relaxed you will feel and help your child feel more relaxed as well!

To Gift or Not?

Definitely not! Parents should not bring a gift for their admission interviewer. While some cultures see this as appropriate, giving a gift, even if it is a small one, is inappropriate and comes with a connotation that is not welcomed at US boarding schools.

STEP 03 EXECUTE

3.1 – UNDERSTANDING THE APPLICATION PROCESS
3.2 - STUDENT AND PARENT APPLICATION
3.3 - RECOMMENDATIONS
3.4 - SUPPLEMENTAL MATERIALS

The success of any good plan is all in its execution. Tim Berry, Founder and Chairman of Palo Alto software company, Bplans.com, and the author of many best-selling business planning books, could not drive this point more when he said, "Good business planning is nine parts execution for every one part strategy."

Managing application requirements and deadlines can be a daunting task and requires an excellent understanding of the application process. One can expect to spend more than 100 hours performing all of the work that needs to be done to complete the application process well and on time. Applying to boarding school is not a task that should be done last minute. The deadlines are absolute — if you miss it by a day, you will not be considered with the current application pool. It's only fair!

3.1 UNDERSTANDING THE APPLICATION PROCESS

Maximizing the application process requires a clear picture of the requirements of the boarding school application and how to use the application systems most efficiently and effectively. Efficiency is important because most students are preparing their boarding school application in the fall. During this time, they are also trying to do their best academically at their current school to prepare for and take SSAT, further develop their X-Factors, and plan and prepare for interviews and school visits. Needless to say, it is a busy time. Students should consider which is the most efficient way for them to complete all of the requirements. Also, efficiency is an important consideration for other people who are going to be helping you prepare your application materials, such as your teachers who will write recommendations and your parents who will complete parent statements.

An effective application takes time to plan, and understanding how the application systems work helps students prepare a compelling application. There are pros and cons to using one system over the other, and knowing which system best helps you portray your brand image is cruicial.

MANAGING APPLICATION REQUIREMENTS AND DEADLINES

The requirements of the boarding school application have been discussed, but let's get everything in one place so you know what you will need in order to apply to boarding school, and where to find it in this book.

Application Checklist

Required

Application Forms/Application Vehicles (Step 3.1)

Student Essays (Step 3.2)

Supplemental Essays (Step 3.2)

Parent Statements/Essays (Step 3.2)

Transcripts (Step 3.1)

Recommendations (Step 3.3)

Graded Writing Sample (Step 3.4)

Application Fees (Step 3.1)

Standardized Testing (Step 1.1)

Character Skills Snapshot (Step 3.2)

Interview (Step 2.3)

Optional

School Visits (Step 2.2 & 2.3).

Multi-Media Links (Step 3.4)

Application Deadlines and Application Fees

Students and parents should check each school's website for application deadlines. Deadlines will also be indicated for each school on the SAO or Gateway to Prep Schools online applications. A few schools have rolling admissions, but generally, boarding school applications are due between January 10 and 15. A few school application deadlines are later, around February 2.

Application fees vary per school and usually differ based on whether the student is a domestic or international applicant. Application fees are listed on the school website and the websites for SAO and GTP.

APPLICATION FORMS AND VEHICLES

There are essentially 3 ways to apply to boarding school. Students are recommended to check each school's website to find information about which methods they accept. The 3 most common **Application Vehicles** are:

1. School Specific Application

2. Standard Application Online (SAO)

3. Gateway to Prep Schools (GTP)

School Specific Application

Some schools do not accept any of the common application methods. Instead, they offer their own School Specific Application. If it is the only method available to apply, then students must complete the School Specific Application. These applications are still online but are hosted by the school or a third-party administrator. The advantage of a school specific application is that it allows students to customize their application to the particular school and student fit. From an administrative point of view, it represents some challenges as it requires recommenders, parents and administrators to complete and send specific materials to that one particular school. Completing a school specific application is a good idea if a student is only applying to that particular school, or if only applying to a few schools and it is clearly the first choice. Generally speaking and referencing back to efficiency, if given an option, students should use one of the standard application methods (SAO or GTP) which are further discussed below.

Standard Application Online (SAO)

The Standard Application Online, commonly referred to as the SAO, standardizes the process of applying to boarding school. The SAO does simplify things a bit because it provides, in one place, all the forms that most school use.

Standard SAO Form List:

Parent Evaluation Form: The need for Social and Emotional Readiness was discussed in Step 1.1. This is a rating checklist to be completed by the parent. Parents are asked to rate their child on certain character skills that boarding school have identified as important for boarding school success. Many are the same skills also evaluated in the Snapshot which will be explained more in Step 3.2.

Parent Statement: This section contains 4 writing prompts for parents to provide information to admissions about their child. There are no word limits on the SAO parent statement.

Student Essay: Contains 4 writing prompts to be completed by the student. There are 3 questions of no more than a 250-word response each and a choice response that is no more than a 500-word response.

Recommendation Forms: Recommendations from teachers and special interest instructors are confidential and should be requested electronically by the student through the SAO application. If using the SAO, it is only necessary for students to request one set of required recommendations. All schools indicated on the SAO will receive this set.

Mandatory SAO Recommendation Forms

English Recommendation Form: A standard recommendation form requesting information about the student's English language knowledge, skills, and personal development to be completed by the student's **current year** English teacher at their current school. To be clear, this cannot be the student's favorite English teacher from a previous year.

Math Recommendation Form: This recommendation form requests information about the student's mathematical knowledge, skills, and personal development and is completed by the student's **current year** mathematics teacher. While most schools will require a math placement test the summer before a student begins classes, the math teacher recommendation also provides information to determine the math placement for the student's first year at boarding school.

Principal/Guidance Teacher Recommendation Form: This is an overall recommendation that relates to all aspects of a student's life at school, including their academics, behavior, citizenship, and contribution to the school community.

Supplementary SAO Recommendation Forms

Personal Recommendation Form: Not all schools require a Personal Recommendation Form. If required, this recommendation addresses personal characteristics of the candidate. Some parents will encourage their child to use this form to find the most prestigious person possible to endorse their child. This strategy is fine if, and only if, the person endorsing has personal experiences with the student and can speak on their behalf. It is preferred to have a meaningful personal recommendation from someone who knows your child well rather than from an "important" person who cannot provide a school more insight into a student's character.

Any Teacher Recommendation: Also, not required by all schools. This recommendation would be from another teacher, other than a math or English teacher. Most often this is requested of a second language teacher or a teacher in an academic area of interest. The Any Teacher Recommendation provides information about the student's subject knowledge, skills, and personal development in the subject area identified.

Official School Report and Transcript Form: This form includes instructions to upload official transcripts and school reports. Most schools require 2 full years of transcripts as well as the most updated grades for the current year. This form cannot be completed by a parent and must be completed by a school official and is requested electronically through the SAO application.

Checklist

SAO provides a vehicle for students to be aware of the status of their application materials under each component of the application. For example, under the Math Recommendation tab, students can track when the math recommendation was requested, when the teacher downloaded it, and when the recommendation was completed and submitted. It is encouraged that you refer to the checklist before calling any school about missing application materials.

Advantages of Using the SAO

1. *The amount of work required to apply to boarding schools can be overwhelming as students also need to keep up with their studies and continue their activities during the application process. Because the SAO is a common application, it does provide students the convenience of applying to multiple schools with one application.*
2. *SAO provides students the opportunity to represent as many activities as they wish and provide in-depth descriptions of each activity.*
3. *SAO allows students to upload as many multi-media links and supplemental documents they wish.*
4. *While students should adhere to the directions regarding character count when completing student essays, most SAO responses have preset character limits that are well beyond the stated requirements giving students a bit of flexibility to include the extra few words or phrases that are essential to their response.*

Disadvantages of Using the SAO

1. *Limitations to customize the application for a particular school is the biggest disadvantage since this is a common application. Students should not make school specific references in any of their application materials.*
2. *Schools that accept the SAO will all ask for a standard set of materials to include: Student Information Form, Student Questionnaire, and Teacher Recommendations (Counselor/Principal, Math and English). However, some schools ask for additional supplements, usually consisting of other student writing responses. The location on the SAO where these requirements are found in the application can be easily overlooked, causing a student's application*

to be incomplete. When using the SAO, required supplemental requirements should be listed under the school's dropdown tab.

Note to Parents

It is worth noting to parents that all SAO Academic Recommendations ask the recommender to comment on whether they have interacted with the student's parents and, if so, if the parent's perceptions of the student's abilities and achievements are compatible with the school's perception of the student. The role of the parent in the boarding school application process has been discussed. Parents should also keep in mind that their interactions with their children's current teachers can impact the student recommendations. Remember back to the situation in the parent interview where a demanding, difficult parent was discussed. Parents who feel that their child's current school is not meeting their student's needs or expectations may want to approach this conversation carefully.

Gateway to Prep Schools (GTP)

Gateway to Prep Schools (GTP) is a hybrid of a common application and a school specific application. It consists of 3 parts: Candidate Profile, School Specific Information, and Recommendations and Transcript Requests.

Part 1: The Candidate Profile

This is the common part of the application. In the section, students provide information that is shared with all schools entered in "My Schools." Information from basic demographics to a listing of a student's extra-curricular activities, academic and personal achievements and awards, and multi-media links are included in the Candidate Profile.

Part 2: School Specific Information

This is the school specific part of the application. Once you identify the schools that you are applying to under the My Schools Tab, you will be able to review the school specific requirements. School specific requirements include essay questions, parent statements, and additional materials such as a writing sample. It is required that students complete and pay for Part 1 before they can complete Part 2. However, it is vital to keep in mind that once you pay for Part 1, there are no longer any opportunities to add or update information in that section. It is possible for students to view the school specific requirements without paying. It is recommended that students work offline and copy the information to the formating required on the GTP only after completing (and reviewing!) that specific section. Only when close to submission, or when the student is confident that there will be no new information to add should payment be made for Part 1. Delaying submission closer to deadlines will provide the most complete application (there is no extra credit given for early submission!). Any updates post submission will need to be sent directly and separately to each school with the hope that it gets placed with the correct application.

Part 3: Recommendations and Transcript Requests

Like SAO, every school on GTP will require a English, Math, and Principal or Counselor Recommendation from current year teachers. Students can request these recommendations electronically from each of their teachers by sending them a link provided in the GTP. Similar to the SAO, teachers complete one set of recommendations which are sent to all GTP schools. However, some schools require additional recommendations, such as a Personal Recommendation, Special Interest Recommendation, or an additional Academic Teacher Recommendation. Some of these other recommendations are required, and some are optional. Students will need to check each schools Part 3 section as it indicates which recommendations are required and which are optional. For those that are optional, students are strongly encouraged to complete them.

Checklist

GTP provides a very handy checklist where students can check on the status of their application materials.

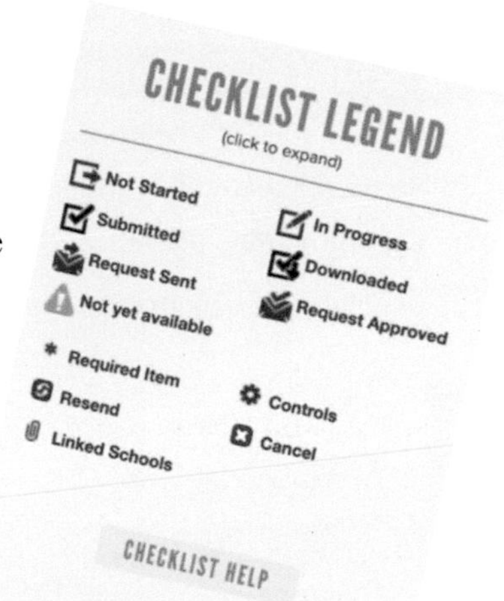

Source: The image taken from the Gateway to Prep Schools website https://www.gatewaytoprepschools.com

Advantages of Using GTP

1. ***Common Part 1 of the application is sent to all schools.***
2. ***Customization in Part 2 is a major advantage. Each school provides its own set of essay questions and parent statement questions, allowing students and parents the opportunity to craft school specific responses.***
3. ***Recommendations do not require a response about parent interactions with current teachers.***

Disadvantages of Using GTP

1. ***Limited response space for students to indicate Interest and Activities, Awards and Achievements, Multi-Media Links and Supplemental Materials.***
2. ***Strict character count limitations.***

Several Different Options Available

Some schools give students a few options on how they can apply. The choice usually comes down to choosing between SAO and GTP. The decision on which to use can be determined by answering the following 3 questions:

1. ***Is a preference expressed by the school?*** *Some schools, while they offer choice, will state a preference. If a school states a preference, then clearly, you should follow their request.*
2. ***Is there a strategic advantage to one application method over the other?*** *Perhaps the student has an excellent response to one essay question on a particular application method or has many activities and multi-media links to upload.*
3. ***Which method is more efficient for the student and family?*** *This is usually the most common reason. Choosing the application method that minimizes time will allow students to focus on other areas such as grades, extra-curricular activities, or test preparation.*

Mix and Match Application Methods

Do not mix and match application methods. If you are applying to a certain school using GTP, you must use GTP for all parts of the application. Setting up accounts for the same school on GTP and SAO can cause duplicate records and confuse schools that you are applying to.

Both SAO and GTP

The reality of the process for most students is that they will likely use both SAO and GTP to fulfill the requirements of the school list that they are applying to. For example, students may apply to 5 schools through SAO and 3 through GTP. However, when possible, do what saves you the most time, but does not jeopardize the quality of your application.

3.2 STUDENT AND PARENT APPLICATION

It is time to put all of your hard work in one place. The student application is the part of the boarding school process where the culmination of Steps 1 and 2 are reflected. Through showcasing activities and awards, writing essays and sharing multi-media links, students have the opportunity for the application committee to get to know their brand and begin thinking about whether or not the school wants to include their brand on the school's shelves.

STUDENT APPLICATION

If students have done an excellent job of defining their brand and their X-Factors, application essays present an opportunity for students to convey the qualities of their brand to their customers who are the boarding schools. The student application should reflect you and your unique voice. Whether the response is to complete a sentence within 150 characters or to write a 500-word essay, each is an opportunity for students to engage an admission officer to learn about them. For the most part, there are no right or wrong responses to the questions asked in the application.

Each admission officer will only have a limited time to read each application, so you want yours to be memorable. Students should approach their application as a way of telling a story about themselves. Try to make your story interesting, compelling the reader to read on. Choose topics that are stimulating and may be different from the writing that you are used to in school. Think about ways to draw the reader to learn more about your brand.

Essays

Application essays are ways for students to demonstrate, and for schools to assess, their writing ability. Essays allow students to:

1. ***Demonstrate academic readiness in written expression;***
2. ***Explain their activities and interests;***
3. ***Express themselves creatively;***
4. ***Share a part of themselves that may not be in another part of their application;***
5. ***Show the person behind the application.***

Keep in mind that most boarding school admission counselors have several benchmarks for comparison of a student's writing ability — they have a student's grades, standardized testing, SSAT writing sample, and recommendations as data points to reference a student's abilities. Boarding school admission officers are adept at identifying inconsistencies in application writing quality. There is nothing that moves a student to the denial pile more quickly than an essay that appears to be written beyond the student's ability. The student must write the student's essays and reflect the student's writing ability and voice.

Students should spend time reflecting on the topics to write about that represent the important areas of their brand and provide more evidence of

their X-Factors. Many schools have several different required responses. Your responses should be well thought out. Students are recommended to gather all of the required applications and reflect on each school's application to identify the general topics or traits you would like to write about in each application set. Try to choose various topics to write about. For example, if you write on the topic of your debate experience for 3 of the 4 responses, you will leave the reader with the impression that your brand will only be available in the "debate aisle." Too much focus on one activity may suggest to the reader that perhaps they should look for a brand that can occupy more aisles of the superstore. Rather, a student whose brand includes an academic passion in physics, demonstrated leadership in community service, and an interest in sports can draft essays or short answer responses that will allow the student to provide stories related to each of these areas. These types of varied topic responses enable the reader to anticipate that this student can contribute to the boarding school in several different ways.

Before getting started on application essays, it may be helpful for students to ask themselves the following questions:

1. *Which event/activity is **most meaningful** to me?*
2. *Which event/activity **highlights good qualities** in me?*

   Commitment, responsibility, cooperation, teamwork, leadership, creativity, resilience etc.
3. *Which event/activity can I **most easily describe or write about**?*
4. *Which event/activity have I **contributed to the most**?*
5. *Which event/activity is **most unique**?*
6. *Which activity has **impacted me the most**?*

All applications require at least one essay; others may require several essays, a combination of essays, short answers, and perhaps a section of fill-in-the-blanks. Whatever the format of the application, students should give thought and spend time composing their responses.

Remember, be sure to follow the direction stated on the application. Most application questions will provide very specific information about the length of essays and responses to questions. The systems (School Specific Application/SAO/GTP) are all online, so they include the spaces between words in the overall character count. Some applications have specific limits. Others give you more room than you need. For example, if a school on the SAO asks students to write an essay between 250 and 500 words, the actual availability for character space provided by the system may be well above the equivalent of 500 words (usually about 3000 characters). Other methods will provide strict character count limit instructions, and additional characters, even one, will cause the response to not be accepted for upload. Some students are tempted to write beyond the directions if the system allows more room. Resist the temptation and stay within the parameters of the instructions. Students need to demonstrate that they are capable of following directions and writing within stated parameters.

Boarding school essay questions vary in many different ways, including:

Length

Long questions

You have just been given the assignment to write your life story from birth until today. What will the title be and why?

Short questions

How have you been helpful to other people?

Question Type

Closed answer questions

If you were to come to XYZ, tell us the one thing you would be most excited about.

Open answer questions

Discuss a situation or issue you were once certain about — but now you are no longer so sure. What changed your mind?

Response style

Critical response questions

If you could solve one global issue at home or abroad, what would it be and why?

Creative response questions

You are back at XYZ for your 20th reunion. Describe your journey since graduation.

When reflecting on your essays, before hitting the submit button, students may want to review the following:

1. Are my essays truthful and represent my own work and voice?

2. Do my essays highlight my good qualities?

3. Are my essays unique and represent my brand?

4. Do my essays demonstrate my writing skills?

5. Do my essays say something about myself that will help the reader relate to me or understand me better?

Supplemental Essays

Why a XYZ School essay?

Some school require supplemental essays. Many schools have a "Why XYZ School" essay, and this presents a perfect opportunity for students to convey why a certain school should include the student's brand in their superstore. If you have specific talents, find out the opportunities available at a particular school for you to contribute and find ways to communicate these reasons in your application. Spend some time on the school's website to find out about some unique programs and activities that fit or interest you. Here is where a

campus visit is also very helpful. Reiterate some of the experiences you had during your campus visit as they pertain to your particular brand and how you may contribute to the school. Perhaps your tour guide shared a specific piece of information that you found helpful, or you met with a certain faculty during your visit. Not all boarding schools are alike. Finding a school that fits you and articulating that fit demonstrate to a school that you have given this process thought and that you are serious about attending their school. Boarding schools strongly consider a student's response to the "Why XYZ School" essay, so take the time to answer this specific question well.

Essays Tips

Here are some tips to help students approach the task of writing essays with confidence.

- ***Read the instructions.*** *If the school requests a page, don't submit five lines or five pages. If answers are to be written by hand, don't type. Some of the online portals will not allow you to submit more than is allowed, but others provide additional space for you to present as much as you have. Follow the directions about word limit — it is representative to the schools of your adherence to instructions.*
- ***Start early.*** *A competitive school list may require 15 plus essays. Writing under the pressure of a deadline only increases the tension and reduces quality.*
- ***Brainstorm.*** *Ask a teacher, tutor, parent, or educational consultant to help the student generate and clarify ideas. Try to think outside of the box.*
- ***Stories are the most effective.*** *Vivid anecdotes that illustrate something about the student's initiative, work ethic, and compassion make the strongest submissions. Try to find an interesting way to deliver your*

story rather than just stating the facts. A "this happened, and then that happened" approach is not very engaging for the reader.

- ***Repurpose ideas.*** *If you're applying to several schools, look for common threads among the essay questions. A strong answer for one school could be adapted to another school's application.*
- ***Proofread carefully.*** *Check for grammar mistakes and logic gaps in your stories. Be very careful when repurposing essays from one school to another. Make sure the essay doesn't say, "I have always wanted to attend X School," if the application is for Y School. This is a deal breaker!*
- ***Authentic.*** *The writing should be that of the student rather than a parent. Parents are certainly encouraged to make suggestions and proofread the final essays, but the writing should be reflective of the student. Schools are making decisions about fit based on variables. One variable is the quality of a student's writing. Elevating essays beyond a student's abilities may land a student in a school that is not a good fit, lead to a great deal of stress and ultimately jeopardize the student's success at the boarding school. Schools will compare application essays with SSAT writing samples. If the quality or level of the two is not similar, students risk being suspected of academic dishonesty and will not be admitted.*

Character Skills Snapshot

As stated, there are many variables that schools consider when deciding who they are going to admit. A student's character matters to boarding schools. The Character Skills Snapshot is an online assessment that usually takes about 30 minutes and is taken at home through the SSAT website. There

Source: The image taken from the SSAT Schools website https://www.ssat.org/snapshot.

is a list of schools that require the Snapshot on the SSAT website. Students can only take the Snapshot one time during the testing year, which runs from August 1 to July 31.

Test experts and school admission professionals designed the Snapshot in recognition that there are skills beyond academics which are important when evaluating students. Skills like **intellectual engagement, teamwork, initiative, resilience, self-control, open-mindedness, and social awareness** are also important data points when schools consider readiness and fit.

Parents need to give permission from their parent portal on the SSAT account for the student to be able to take the Snapshot. Once permission is given, the Snapshot becomes available to the student through the student portal on the SSAT account.

The Character Skill Snapshot captures a moment in time and is not an absolute measure of a student's character. It is expected that students will continually develop these skills throughout their lives.

Like the SSAT, the Snapshot is a normed test, and the results are based on a comparison of similar age students. Students are evaluated on each of the 7 Character Skills mentioned above. Their performance is broken down in to 3 categories: Emerging/Developing/Demonstrating.

1. ***Emerging category:*** *The student scores below the 25th percentile when compared to his/her peer group, indicating that the student is starting to show signs of this skill.*
2. ***Developing category:*** *The student scores between the 25th and 75th percentiles when compared to his/her peer group, meaning that the*

student in the norm group displays this skill, but the skill is continuing to develop.

3. ***Demonstrating category:*** *The student scores above the 75th percentile when compared to his/her peer group, and the student scoring in this category displays a clear understanding and use of the skill. This does not mean mastery. There is still room to grow. Generally, one does not expect a middle school/early high school student to be demonstrating all 7 of these characteristics.*

The introduction of the Snapshot has been a useful benchmark for helping students to recognize the importance of preparing for the journey to boarding school and ultimate success if admitted. The emphasis on these characteristics conveys that schools are looking for students with at least some of these skills. Like academic readiness being demonstrated through grades, SSAT scores and recommendations, social and emotional readiness can be demonstrated through character. While initially, boarding school may seem like a "fun" experience, a little like summer camp, the newness and novelty fade and boarding schools are ultimately academic institutions. There are times at boarding school when all students struggle, whether in academics, residential life, or some other aspect of boarding school life. Having some character skill development for self-control, social awareness, and resilience can be very beneficial to help work through the inevitable challenges of boarding school life.

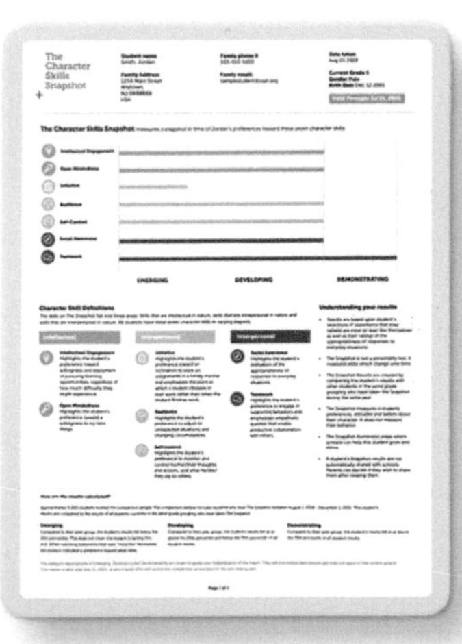
The Character Skills Snapshot

Source: The image taken from the SSAT website https://www.ssat.org/snapshot.

Boarding schools are founded on the principles of intellectual engagement, teamwork, and open-mindedness. Students who don't at least have some of these skills at the onset of attending boarding school are not likely to engage in boarding school life. As students look to prepare themselves for boarding school, reflecting on these skills is a good exercise. Like you would if your SSAT scores needed improvement, identify your character skill weaknesses and seek opportunities to further develop in these areas. This preparation will be one of the most valuable for your ultimate success in boarding school and life.

PARENT APPLICATION

The summer before your child is applying to boarding school is an excellent time to start thinking about your parent application or parent statement. Getting a draft together over the summer will save you time later and help prepare for your parent interview during school visits.

Many parents ask about the purpose of parent statements and whether they matter. The purpose of the parent statement is to add a dimension to the candidate statement and help the admission committee better understand the applicant from a parent's perspective. Parent statements are also used by schools to help identify whether parents are a good fit for their school philosophy and culture. Admission officers are also looking for indicators in parent statements to answer some important questions like the following.

1. ***Are these particular parents likely going to be partners in the child's education, or will they be too demanding? Do the parents seem like they would be reasonable if their child has any difficulty?***
2. ***Will they represent the values of our school within their local communities?***
3. ***Will they participate in our school community?***

Parent statements are the parents' chance to provide a personal introduction to your child by sharing details about how your child learns, and his/her interests and strengths. While most parent statements will neither admit nor deny a student, boarding schools are communities and looking for family members (e.g. parents) who are aligned with the school's philosophy and will be present within the community.

Steps to Writing an Effective Parent Statement

Start with SAO

Most schools require parent statements. SAO schools have 4 parent essay questions required by all schools. For schools that accept the GTP application, like the student essays, the GTP parent statement questions vary for each school. Most students have schools on their list that require SAO. It is most efficient for parents to brainstorm and complete the SAO parent statement questions first and then look at the requirements of the GTP parent statements to see if any responses or content can be "recycled " and then customized to fit the particular GTP school.

Think About Your Responses

Understandably, most parents have difficulty in objectively evaluating their child. During my initial meetings with families, parents beam with pride as they boast of their child's accomplishments. It can be difficult for parents to step back and consider their child objectively, but you need to do precisely that. Parents may want to reread report cards and teacher comments over the past 2 years. Do you see any consistent themes that emerge from the reports? Comments from teachers provide valuable information about how your child learns and acts in school and in extra-curricular activities. Consider also your observations of your child, as well as what you hope your child will get out of his/her boarding school experience, not only in academics.

Be Honest

Real children are not perfect, but they can still be great candidates for boarding school. Admission officers are interested in areas where your child may have been challenged or even failed. Describe your child accurately and openly. A full, real, and descriptive parent statement will convince the admission committee that you are being honest, and will help committee members understand your child and what he/she can offer a school.

Children who succeed in their schools are happier and healthier and stand in better stead for college admissions. Parent statements also provide the opportunity for you to describe your child's strengths. You should not feel the need to be negative — but everything you write should be truthful. Avoid duplication of your child's CV or listing awards and accomplishments as the primary content of your parent statement. There is nothing more of a turnoff to admissions than a parent brag sheet. Rest assured, if your child has done their job executing the plan described in this book, their achievements and attributes are represented in their application. Instead, try to provide stories that support your child's accomplishments or identify personality strengths that are desirable to boarding schools like resilience, teamwork, and intellectual curiosity, to name a few. Remember the Character Skills Snapshot. These are the skills that boarding schools are looking for in students. Your child does not need to have demonstrated all of them but choose a few that you feel reflect your child. Provide some antidotes or examples of how these skills are expressed in your child's experiences.

Consider How Your Child Learns

The Parent's Statement is a chance to describe how your child learns so that the admission committee can decide if he/she is likely to take advantage of the opportunities at a particular school. If your child has moderate to severe learning issues, there are areas of the parent statement that ask you to provide this information. Many private schools grant students with learning issues accommodations or modifications in the curriculum so that they can best demonstrate what they know. Some schools have specific programs and resources for students with learning differences.

Parents of children with a learning difference are encouraged to research what kind of resources the school has available to help their child before he/she attends the school. If your child has had an educational evaluation, speak with the professional that administered the battery of tests, and discuss your child's future needs if they are to attend boarding school. Being open and honest with the school will help you and your child find the school where he/she can be happy and successful.

Students who find themselves admitted to schools that are not a good academic fit are at risk of not being successful and returning home. For some international students, returning to school in their home country can be difficult once they leave. Some families of students who have returned home find that they are no longer able to return to a previous school and can end up enrolled at a far less competitive school in their home city. Returning home from boarding school is a no-win situation, so finding a school that best suits your child will help ensure long term succcss.

Tips for Writing the Parent Statement

The best parent statements tell stories about their children. Try to be creative and provide as many anecdotes about your child as you can. When brainstorming, parents are encouraged to give examples of specific attributes or personality traits to add color and context to their statements. Feel free to use humor. Consider the admission officer who will read your parent statement. Is your parent statement compelling enough to be memorable?

These are some of the questions that may help parents brainstorm their parent statements. The following questions will help parents to respond to most of the questions they will encounter across all school application parent statements.

1. ***What are the reasons you are considering an independent school for your child?***
2. ***What do you believe your child might contribute to a school community?***
3. ***Are there particular skills or talents you hope will be cultivated at your child's next school?***
4. ***What are three adjectives to describe your child? Please write a paragraph explaining why you feel this represents your child. Tell us a story about your child and how this adjective applies.***

5. ***What qualities of character and mind in your daughter or son most delight you?***
6. ***What do you see as your child's strengths and weaknesses?***
7. ***Do you have any concerns about your child's readiness for an independent school?***
8. ***What has posed the biggest academic and/or extra-curricular challenge for your child? How did/has your child responded to this challenge?***
9. ***What role does your child play in your family? How do they interact with siblings, grandparents, mother, and father etc.?***
10. ***What was your child's proudest moment?***
11. ***Where do you see potential growth for your child?***
12. ***What are your educational goals for your child?***
13. ***Tell a funny story about your child.***
14. ***What is your parenting philosophy?***
15. ***Is there anything about the sequence of your child's schooling that we should know? Did your son or daughter ever skip or repeat a year? Was your son or daughter ever asked to withdraw from a school, suspended or put on probation?***

3.3 RECOMMENDATIONS

Recommendations are a critical component of the boarding school application. Recommendations provide the admission committee an objective review of how your brand has performed in their establishment.

American boarding schools are seeking students who are successful in the classroom, demonstrate intellectual curiosity, engage in their educational process by asking questions, think critically, and collaborate in problem-solving. Schools are looking for indicators of these qualities in recommendations.

Recommendations usually have a rating component where students are evaluated on different variables as compared to their peers. Ratings range from questions about willingness to take intellectual risks, open-mindedness to the ideas and opinions of others, as well as specific skills required. Recommenders are required to respond to questions like:

1. ***Does the student raise his/her hand, contribute to classroom discussion, and ask questions?***
2. ***Are they a student who does well, but remains quiet in the classroom?***
3. ***Do they demonstrate leadership in group projects, or take over and do not collaborate with their peers?***
4. ***Is the student disruptive in the classroom?***

Like most everything else in this process, learning and demonstrating the desirable qualities takes time, particularly if a student's current school does not typically reward some of these types of characteristics.

The amount of time for teachers and staff to complete the recommendation requirements for students to apply out is sizeable. Most schools do not want to lose their students to boarding school and, naturally, can be less than enthusiastic about the requirements asked of them to support a student who may be leaving to attend another school. Still others have their procedures and policies which may not comply with the existing boarding school application methods. Students and parents may find themselves needing to be creative in how they approach recommendations from their particular school.

Since recommendations are requested in the fall of the application year, parents express concern that the current teacher has not had enough time to truly get to know their child to write a favorable recommendation, or perhaps the child had a "better" relationship with last year's teacher. The requirement remains: academic recommendations must be from the current year teacher. In middle schools in the US, teachers usually change each year. Since teachers will write the recommendations with no more than 3–4 months of experience with the student, most students are on equal footing with their current year teachers, and boarding schools recognize this fact.

Recommendations must be communicated directly from the teacher/school to the boarding school through the recommendation portals on the SAO/GTP/School Specific Application. The chain of custody of recommendations provides teachers a vehicle to share confidential feedback about students. Parents should not send recommendations. The recommendation requirements were defined in Step 3.1. In the section below, specific responses asked of recommenders will be discussed to help students to understand and recognize the types of readiness indicators that recommendations can supply.

ENGLISH RECOMMENDATION

The recommendation from a student's current English teacher is very important as English proficiency is a critical variable of fit, as was mentioned in Step 1.1. English teachers are asked to respond to questions such as:

1. How does the student's overall performance compare to that of other students in the grade level?

2. How accurately does the student understand the material that has been read?

All recommendations also ask for information that relates to a student's character and learning style:

1. Please comment on the student's character, citizenship, and contributions to your community in comparison to other students in this grade level.

2. How well does the student accept advice or criticism?

As mentioned previously, the SAO recommendations ask teachers about parents:

1. Have you had an opportunity to interact with the student's parents/ guardians? If yes, was the parent's perception of the student compatible with your school's perception of the student?

MATH RECOMMENDATION

Math teachers rate students on mastery of topics related to math such as Pre-Algebra, Algebra I and II to Calculus BC. In addition to computation skills, math teachers rate their students on leadership potential, willingness to accept the challenge of more difficult problems and exercises, and also on creativity.

Most schools will send enrolled students a math placement test the summer before they enter boarding school. However, the math recommendation, in addition to commenting on specific quantitative skills, requires teachers to provide an endorsement about math placement for the student's first year at boarding school.

PRINCIPAL/GUIDANCE COUNSELOR RECOMMENDATION

This recommendation is the most general and asks about student characteristics, school attendance history, and general contribution to the school community. Principal/guidance counselor recommendation refers to questions such as:

1. How does the student's overall performance compare to other students in the grade? Please elaborate.

2. What three words come to mind when describing this student?

3. To your knowledge, are the parent/guardian's perceptions of the student compatible with your school's perceptions of the student?

4. Please provide any additional information that will give us a more complete picture of this student.

SPECIAL INTEREST RECOMMENDATION

This recommendation is likely going to describe an area that would be considered an X-Factor. Special interest recommenders can be affiliated with the school or can be outside coaches, music teachers, community service directors, or Model United Nations teacher etc. The special interest recommendation asks the recommender to comment on the specific skills or area of interest. Special interest recommendations are not required to be from a current year recommender.

PERSONAL RECOMMENDATION

The personal recommendation is the recommendation that families have the most difficulty in identifying the recommender. By nature, a personal recommendation needs to be written by a person that has had direct experience with this student over time. For some students, the person to identify could be someone from their church, a boy scout leader, or the faculty at a community service program that the student has volunteered at over time. This person's experiences with the student demonstrate qualities such as their character, resilience, and how they deal with challenges.

OTHER RECOMMENDATIONS

Parents sometimes ask about whether or not recommendations from "important" people affiliated with a school can help. The answer is maybe. A person of influence may be a current parent, a member of the board of trustees, or an alumni. These recommendations can be helpful because they are coming from an individual that knows the school and, because the recommender knows the student and family, he or she can assess the student as a fit. When considering these types of recommendations, the recommender should meet with the student. If you find that you do have some connected support at a specific school, but the recommender does not already know your child, please arrange for them to meet so that the recommender can learn about your child. It is also helpful for your child to meet someone else from the school. This way, if the recommender does write to the admission committee on your child's behalf, he or she will have direct experience to recommend your child's brand, and your child can also ask the recommender questions about the school. In marketing terms, this is like a celebrity product endorsement!

TIPS FOR EARNING THE BEST RECOMMENDATIONS

Academic recommendations provide schools with information from your teachers about how well you are doing in the classroom, about your general performance, and how you behave in the classroom as it relates to motivation towards learning and interacting with your teacher and peers. Recommendations can make or break your application, and good teacher recommendations are vital to gain entrance to boarding school.

The Behavior

Like everything else, start early.

- ***Establish personal relationships.*** *A letter of recommendation wouldn't be great if the person writing it doesn't know you well. Make an active effort to develop meaningful relationships with your teachers and counselor! Participate in classroom discussions. Ask your educators intelligent questions after class. On a day when you feel that the class was particularly interesting, tell your teacher so. Reach out to them during office hours. If you find a particular assignment enjoyable, say so when you hand it in. When you consistently demonstrate passion both inside and outside the classroom, your teachers and counselor will be able to produce a strong personal endorsement based on the positive character traits that you have demonstrated through these efforts.*

- ***Everyday behavior matters.*** *When asked to write a recommendation, teachers and counselors draw from their daily encounters with you, so exercise awareness in how you behave with your teachers. Greet your teachers in the hallway every morning, use respectful language in your emails, smile a lot, and try to ask a question a day in class. Needless to say, teachers will also remember you if you are always late to class or if you had a big fight with your classmates over a project or assignment. The goal is to make a positive impression, rather than a negative one, through mindful communication and behavior.*

- ***No gifts to teachers.*** *While sharing a cookie with your teacher may be seen as a gesture of kindness, giving him/her extravagant gifts will raise some eyebrows as to what ulterior motives you have. Ultimately, it will reflect badly on you. In general, stay away from offering gifts to your teachers and counselor. Instead, show appreciation for their effort and guidance through smiles and a heartfelt "thank you" note from time to time — this will be very much appreciated!*

The Ask

By November 1st of the year you are applying, ask your recommenders to write a recommendation for you. Your teachers will respect you if you do this personally and in person rather than having your parents ask or sending an email request. Most adults will value the character that you display if you ask them face to face.

The Details

Reflective of the school list, students should create an instruction list of which schools require which recommendations and instructions for the recommenders. You will need to inform your recommender of the method that they will use to complete the recommendations — either through the SAO link that you will send them, or the GTP link you will send them. For other schools, you may have to send them a link to complete the recommendation from a school specific application site. If the school does not comply with the online systems, you may also need to provide your teachers with the general admission email of the boarding school where they will send the recommendation directly. While it can be complicated, try to make it as easy as possible for your recommenders.

The Follow-Up

Please be sure to follow up to confirm the submission of your recommendations well before the deadline. For those recommendations that will be submitted through SAO or GTP, the online systems allow you to track the status of the recommendations. For recommendations that are sent directly to the schools, you will need to check the school's application portals or contact the schools

directly. After the application deadline, most schools will send an email informing students of any missing items. It is the student's responsibility to follow up on missing application materials. Don't forget to say "thank you" to your recommenders!

SPECIAL CONSIDERATIONS

Additional Recommendations

Please keep in mind that students should not request more than 1 or 2 additional recommendations, beyond the stated required recommendations. More is not always better. Admission officers are very busy, and additional recommendations should provide information that is essential to the application process. Additional recommendations outside of the SAO or GTP need to be sent directly to schools, which can be confusing for the recommenders. It is crucial to provide very specific information to the recommenders about how, when and where to send additional recommendations. It is best to provide the general admissions email to each additional recommender and ask them to email the recommendation directly from the school with the student's full name in the subject line of the email.

3.4 SUPPLEMENTAL MATERIALS

Boarding schools encourage students to submit materials that will supplement a student's application. By definition, the word "supplement" means to complete or enhance something else when added to it. Effective supplemental materials add to a boarding school application, rather than detract from it by not being current, significant or compelling. There are several areas of the application that provide opportunities to add meaningful materials.

MULTI-MEDIA SUBMISSIONS

Multi-media links can be thought of as the show and tell of the application, and they usually relate to your X-Factors. If you are a talented musician or athlete, then a link to a video of you demonstrating your talents can be beneficial to coaches and admissions in evaluating your skills in this area. Examples of multi-media links that students have provided in the past include YouTube videos, music files, art portfolios, and presentations. Be sure to provide a well-written description of the content in the space provided. Many schools no longer accept CDs, DVDs, videotapes, or hard copy portfolios. Links are the simplest and most commonly accepted way to send multi-media to schools. The SAO application does not have any limits to the number of multi-media links that students can send, while GTP schools allow students to

submit 2 multi-media links in the Multi-Media Link Section and no more than 3 more links in the Additional Information Section.

Be sure to be mindful of an admission officer's time when sending multi-media links. Multi-media links should be no more than a few minutes in length. If there are other people in the link such as in a soccer game or drama performance, help the admission officer identify you by adding an arrow to yourself or some other identifier like a description of what you are wearing. Don't forget to include passwords in your descriptions if the links are private. Make sure what you send is current and relevant. While your parents may think that it may be a good idea to send a video link of you playing the violin when you were 5 years old, or your first "masterpiece" painting at 10 years old, admission officers only want to see your recent (a two-year window is a good guide) and your best work. If you send too much information, admission fatigue will set in, and chances are, the admission officer will not be impressed.

GRADED WRITING SAMPLE

Not all schools require a graded writing sample. However, most students are likely to have at least one or two schools on their list that require them to submit a graded writing sample. It is prudent to know the purpose and requirements of a graded writing sample in regard to planning ahead. Graded writing samples are important because they show a school not only a student's writing skills but what type of work you are doing in your current classroom, expectations from the teacher, and how you are performing. The requirements of a graded writing sample can vary, but most require that the writing is from a current year English or Humanities class. This sample must contain grades, which means the sample has some teacher's comments and evaluation indicated on the example. Many schools have online grading systems or Google Docs, which doesn't necessarily allow for what is thought of as the old-fashioned way where teachers provide handwritten comments within the margins of the paper.

Depending on how your school grades papers, you may need to get creative with technology to meet the requirements of the graded writing sample upload. We suggest that if a student's writing sample is only available online, take a screenshot of your teacher's comments and insert them on the appropriate document, then save that document as a PDF file for uploading on the application systems. Graded writing samples should be sent to the school directly by the student within the application method the student is using.

Many international schools and schools abroad don't require students to write as much as American schools. Some students have difficulty finding an appropriate writing sample for submission. Students also realize this requirement last minute and scramble to identify a sample for submission. When you start school the fall of your application year, be aware that you will likely need a graded writing sample. Take a look at a syllabus or ask your teacher about any upcoming major writing assignments. Make sure to do your best work. Despite poetry being a section on the SSAT, most schools do not want students to submit poetry as their graded writing sample.

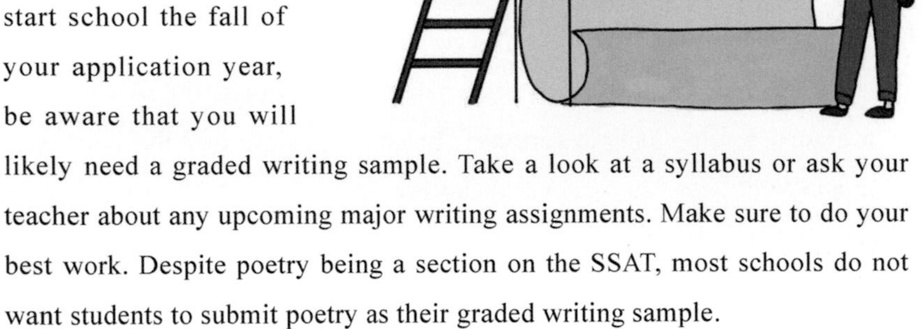

ADDITIONAL MATERIALS

Additional application materials could be just about anything that you feel would enhance or support your application. Things like an article that you wrote for your school newspaper or a writing contest, or a document of code that you wrote when you developed your first App would be good examples of additional materials. Try to find "something" meaningful to include here rather than leaving the section blank.

Most schools do not like students to provide them with hard copies of application materials. Schools are not only trying to promote sustainable practices, but also responding to the potential volume of hard copies that would need to be cataloged and entered into student records and files.

There are certainly ample areas on the SAO for all materials that you could want to submit. However, the GTP system provides fewer opportunities to add additional materials. Some GTP schools give students an additional opportunity to attach a PDF file of additional supplemental application materials. Try your best to make sure you have entries if given the opportunity to provide additional information.

SPECIAL INTEREST CURRICULUM VITAE

If a special interest is a prominent X-Factor, students may want to consider creating a curriculum vitae (CV) or resume specifically about their experiences and achievements in this particular area. This is a concise document that affords the reader the opportunity to see all interest-specific achievements and experiences in one place. Creating the content for a CV takes time, and while many students begin their interest in these areas when they are quite young, we recommend that only recent experiences and activities are included (past 2 years). One can indicate on the CV that the student has, for example, been playing tennis since they were 5 years old, and taking regular lessons. However, including specific achievements and activities related to tennis for more than 2 years before the application year would not be recommended.

STEP 04 FOLLOW-UP

Communication builds strong relationships and good follow-up requires consistent communication. Step 4 is all about the importance of follow-up in the boarding school admissions process. Follow-up provides an opportunity for students to reach out to their loyal customers to remind them of their brand, provide new information and answer any questions. Follow-up is an excellent opportunity for students to continue to show how and why your brand should be chosen. The key to effective follow-up is to make sure your interactions have added value, even after applications have been submitted.

4.1 INCREASE VISIBILITY

It would seem to make sense that the submission of the application marks the end of the application process. Families can finally go ahead, sit back, and take a break until decisions are released. Not so fast.

While the heavy lifting of completing the application is indeed in the past, there is still work to be done to make sure that you continue to build your loyal customers who want your brand front and center on their shelf. It's all about **Increasing Visibility** during the follow-up stage of the process.

If you have followed the recommendations in this book, you have established an open communication line with your admission officer and other faculty at the schools you are applying to. As always, when communicating with your customers, you don't want to waste their time. You want each interaction to be meaningful and contribute to your brand recognition and strength.

Most applications are due early January, and admission decisions are released around March 10th. The time between submission and decision allows you to continue to build your brand at the schools on your list, while the schools are beginning the process of deciding which products to put on their shelves. Like any business pitch, whether it be a new phone plan or an application to a prestigious boarding school, Follow-Up with the customer is essential.

Think about the things that you have done since you submitted your application. Most students can find something interesting in between the submission and application decisions. Have you had any experiences that were particularly meaningful or had any achievements or improvements in one of your X-Factors? Perhaps you have discovered a new interest or had a unique experience.

Admission offices encourage students to send additional materials that will provide further support for their candidacy. The key is to give some information that is significant enough for the admission office to take the time to read and add the information to your admission folder. Admissions updates should be sent in an email to the admission officer you met directly during your school visit. If you found throughout the process that your admission officer has not been responsive, you may want to consider copying the Director of Admission and the general admission email as well. Lack of communication with you is not the sign of a loyal customer.

ON-GOING COMMUNICATIONS WITH LOYAL CUSTOMERS

If you have followed the recommendations made throughout this book, you will have reached out to key staff or faculty at the boarding schools. You met with them during your school visit to build your brand and encouraged them to join your loyal customer base. Your brand has not been bought by the customer yet, so you still need people in your camp who believe that you can succeed at the school. As a follow-up, provide your loyal customers with an update on the specific area you can contribute at the boarding school. If you have been in touch with the Orchestra Director at the school of your choice and learn post-submission that you have been chosen for a select orchestra, write to them and let them know. If you have new materials to send, such as an updated multi-media link for art or debate, go ahead and send it. If swimming is one of your X-Factors and you have been in touch with the swim coach, let them know you recently earned a personal best, including the event and time. Also copy your individual admission officer on these communications with your other loyal customers. It's important that everyone is aware of your excellent follow-up.

You have built your customer loyalty so that the schools can trust your brand. You have applied to a balanced list of schools. Hopefully, throughout this process, some schools have emerged as a better fit than others. You now have a better understanding of yourself and the boarding school process. You can identify your strengths and weaknesses and understand how they relate to your application and chances of being admitted.

Generally, there are more qualified applicants than there are available spots at US boarding schools. When making admission decisions, schools have to carefully try to predict the number of students who will attend their school if accepted. This is commonly referred to as "yield."

Schools want to be as close to their yield projections as possible. If too few students accept their offer of admission, schools will be left with empty spots and will need to scramble to fill the vacancies. A low yield may also affect future selectivity ratings. The opposite scenario occurs when more students accept the offers (i.e., yield) than expected. 'Over-yield' creates a situation where the school could be over-enrolled — stretching resources at the school and impacting the following year's numbers as larger classes advance a grade the next year.

Demonstrating interest throughout the application process is an important fit factor for most schools. If you can narrow down your school preferences to a top few, it may be helpful to let them know this. Here is where communicating and demonstrating your preference for a school may help your chances of being admitted. Let's put this in context by discussing a scenario between two equally qualified students.

Candidate 1 has been very communicative throughout the process. Candidate 1 has interacted with various faculty and staff outside of the admissions office and has updated them throughout the process. He/she has demonstrated fit to the admission officer with whom he/she has been working. The applicant is a good academic fit for the school, and several of the student's X-Factors stand out at this particular school. The student likes the culture of the school and the people that he/she has interacted with. Candidate 1 has expressed his/her strong interest in attending the school.

Candidate 2 is also academically in line with the school expectations, and this student's X-Factors also have an impact. Candidate 2 has done very little follow-up after submitting his/her application.

Admissions Perception: Candidate 1's brand is well represented at the school, he/she has built a solid loyal customer base, and the school could easily place his/her brand on their shelf. Candidate 2's brand, while it may be equally good, doesn't provide as much confidence to the customer — the boarding school. In most cases, Candidate 1 will be more likely to gain admission than Candidate 2.

If you cannot identify a few preferred schools, it is best to pursue every fit school as if it is your top school. However, unless it is your first choice, you should not communicate the first-choice preference to any school.

Students need to understand the competitive landscape of US boarding school admissions and their chances of being admitted to certain schools. If you started with a balanced list, getting an offer from a school on your list will be more likely. There are fantastic schools at all different levels of selectivity. Make sure that you give attention to the schools on your list that you may consider a good fit, albeit less competitive. It would be best if you had schools on your list in all categories that fit you. Students who do not heed this advice and only apply to top schools may find that they are left with no offers after all of the hard work and time devoted to the application process. In this scenario, your brand will, unfortunately, need to remain on the shelf of your current school.

4.2 THE ADMISSION DECISION

Most schools will release **Admission Decisions** on March 10th. Students and parents should make sure to check email in advance to be aware of how a school will communicate the results. Some schools send emails to students with the admission decision. Some schools post the decision in the student's account, while other schools may send the admission decision via postal services or even make a personal phone call. March 10th is stressful enough without having to figure out where to get the news!

As discussed in Step 2.1, with a balanced list, there will be "reach schools" that are highly selective. Reach schools refer to schools that are slightly more competitive than the student's profile, but still represent a potential fit. Hopefully, these schools are not so far of a reach that the student did not ever have a chance of gaining any loyal customers at the school. There should be some "target schools" on the list that are still selective, but the student matches the school's profile and entry requirements. A balanced list should also include other schools that the student may be slightly above the school's profile, but the school still represents a fit for the student. Those schools are known as "likely schools." Mostly, there are 3 outcomes to the boarding school application process: **Accept, Deny, or Waitlist.**

ACCEPT

During the admission process, it has been the family's mission to get their brand on the shelf of the superstore. Students and parents have worked to develop, define, and market the brand. Through this journey, students have gathered loyal customers and ultimately sold their brand to the boarding school.

Congratulations! Gaining admission is the best news and the desired outcome for all students in this process. An **Accept** offer means that the student now needs to identify into which superstore they are going to place their brand. It is time to decide on which boarding school to enroll. You have until April 10th to decide.

One and Done

The "One and Done" student received one offer of admission, was not placed on any waitlists and plans to enroll in the school from which they received the offer. Students in this situation should complete any electronic enrollment confirmations and draft an enthusiastic letter to their admission officer, letting them know of their decision to enroll. Be sure to thank the admission officer for their support throughout the process (brand loyalty).

Multiple Offers

Some students will find that they have several offers of admission. If students followed the balanced school list recommendation, they may find that they have offers from schools that they like and offers from other schools that would be less desirable or considered less of a good fit. If a student was admitted and knows that they will not be attending School X, then it is a courtesy to let the admission officer know sooner, rather than later. There is likely a student on the waitlist of School X that would be delighted with the vacancy created by the declined offer. For those schools that a student is not considering attending, students should compose a very polite email to their direct admission officer declining the offer of admission and complete any automated responses required to decline the offer. Students should thank the admission officers for the confidence placed in them when invited to attend the school.

Denial is a difficult result with any school list. The admission committee has given careful consideration to every candidate's application, and a denial for whatever reason marks the end of the process at that particular school. Schools will not identify or state a reason for a denial. Everyone deserves a chance to dream, and even a denial at a school that the student has very little chance of being admitted can be disappointing.

Usually, there is no further communication with a school after a **Denial**. However, if, in the end, the student receives no offers and plans to reapply, then it would be suggested that an email be sent to the student's admission officer thanking them for the opportunity to apply and wishing them the best. **Re-Application** will be addressed later under "Special Considerations." A polite reply following a denial of admission will finalize the current relationship and provide a conduit for the student to reconnect later if needed. Additionally, students may have younger siblings who will apply in the future, and it is always a good idea to close a relationship on a professional note.

While a denial decision is difficult, it is short and complete. A denial indicates that a school does not want your brand on the shelf; you need to move on. A waitlist decision, while offering a glimpse of hope, can also be described as painful, like a dull ache that can last a long time. A waitlist decision indicates that a school likes your brand well enough, but there is no room right now on the shelf of the superstore.

The waitlist relates to Step 4.1 regarding post-submission follow-up and schools trying to predict yield. Most schools will have some sort of waitlist for qualified students. The number of students on the waitlist is a factor of the school's yield calculations and past history. In the past few years, there has been a shift in the waitlist philosophy of some schools, who now are denying more students rather than having long waitlists. This shift frees the student to bond with a school where they were offered a spot on the shelf or they may come to the realization that they may want to strengthen their brand in order to reapply the following cycle.

Waitlist Communications

In the waitlist letter, schools will provide very specific instructions on how students should respond to the waitlist offer. Generally, the student has the following options:

Option 1: "No, I do not wish to remain on the waitlist."

Follow the school's instructions provided in the waitlist letter. If a student has had consistent interactions with their admission officer, it is a nice touch to send the admission officer a polite thank you for their support and let them know where you decided to enroll. Most schools track enrollment information of students who decline their admission offers and waitlists. Think of the superstore model — this is a way for the boarding schools to track their "competition."

Option 2: "Yes, I would like to remain on the waitlist."

Asking to remain on a waitlist implies that you are still interested in attending the school. Students should communicate their desire to stay on the waitlist only at schools that are more preferred over schools that they currently have an offer from or if the student has no offers. Since waitlisted students are not invited to revisit days, waitlisted students will not have any additional data points to inform a decision about accepting a waitlist offer. Remaining on a waiting list to see "if you can get an offer" is a waste of your time and admissions' time unless you prefer the waitlisted school over the school(s) at which you have an offer.

Follow the protocol identified by the school to remain on the waitlist. This usually involves some sort of electronic indication of your intent. In addition, students should reach out again to their admission officer, who hopefully is a loyal customer, albeit a bit reluctant, hence the waitlist. Compose a thoughtful email to the admission officer. Thank them for the opportunity to be on the waitlist. Remind them of your brand and the X-Factors that fit their school. Provide any new information since your last update. Then it is time to wait! After following the recommendations in Option 2, you should not continue to follow up with the school unless instructed to do so. If the school wants to offer a spot to a particular waitlisted student, they will be in touch. Movement off the waitlist doesn't often happen with international students from competitive regions or regions where there are many applications. Prolonged time on the waitlist can be difficult for students to manage. As long as students hold out "hope," it prevents students from bonding with the school that they enrolled at or committing to alternative arrangements. Students should give thoughtful consideration to what it means to remain on a waitlist for a prolonged time.

Most schools will eventually send all applicants on the waitlist a communication that the school is no longer considering applicants from the waitlist. Now it is time to get ready and fully embrace the school that you have enrolled at or make other plans.

Parents are reminded that they must enroll and deposit at a school by the date specified in the acceptance offer. If there is movement off the waitlist, it does not usually happen until sometime well after April 10th when all students are required to notify schools of their decision to enroll or not. Keep in mind that if your child is offered a spot from another school's waitlist and chooses to attend that particular school, any deposit or tuition payments that you have made to a prior school will be forfeited.

THE DECISION

Some families start off the boarding school process convinced of their school preferences from the get-go. They base their decisions on perhaps having attended a summer program at the school, on friend's recommendation, or quite often on published rankings. After going on school visits, students either further refine their preferences or form new opinions and communicate their choices to the school during the admission process, as discussed in Step 4.1.

For students who have multiple admission offers, resist the temptation to decide where to enroll based only on your admission experience to date, school rankings, or where friends are attending. For some students, the decision represents the next 4 years of their lives, at least USD250,000, and the foundation of their education in forming their university/college options. Needless to say, it's a big decision!

Think about the amount of time and interactions from which families are making this critical decision. Let's count the average time one might have direct experience with a school during the admission process.

- *Visiting school websites: 1-2 hours*
- *School visit: 2 hours*
- *Attending local reception: 1 hour*
- *Emails with admissions: 1 hour in total*
- *Emails with coaches and various faculty: 1-2 hours*
- *Networking with friends: 1 hour*

Let's add it up and round up by adding one hour. At best, most students may have spent 10 hours interacting with a school! Truly, not a lot of time from which to base this decision.

Revisit Days

Revisit days represent the most in-depth look a student will have before making the decision. A few schools offer overnight visits where students can also experience what happens outside of the classroom. All schools offer admitted students the opportunity to revisit the campus before deciding to enroll. Ideally, all admitted students should attend a revisit day at every school they are considering attending. Revisit days can be one of the most effective methods for families to determine and confirm fit. Most schools offer a few options for revisit days, which are typically scheduled the week leading up to April 10th. Your acceptance letter should identify the dates of the revisit days or alert you to where to find the information and how to register. Waitlisted students are not invited to revisit the school.

Revisit days are very different from admission visits and tours. There is a shift in customer dynamics. After an acceptance offer, the student becomes the customer, and the boarding school is trying to encourage the student to be part of their superstore. During a revisit day, families are usually invited to "try out" the school. Revisit days are for both students and parents and typically involve attending a class or two and various interactions and activities. There may be panels for students and parents where current students share their experiences and answer questions. Schools may showcase the talents of their students through drama and musical performances. Revisit days are designed to show off the best of a school so that students will choose that school.

Some international families are reluctant to attend revisit days or do not recognize the importance of completing this last step. Understandably, attending a revisit day represents additional planning and resources dedicated to attending. Traveling to the United States again may require students to miss school and parents to adjust their schedules. Years of experience shows that every year at least one family who is convinced that School X is the best fit, selects School Y after attending a revisit day at both schools. It is a lot to ask, but the time and resources will be well spent on investing in attending revisit days.

One and Done

If a student knows which superstore their brand would thrive in, why should they need to attend a revisit day? Attending a revisit day can be beneficial to transition to boarding school. Admitted students will meet other admitted students. Connecting with future classmates and making a few acquaintances

can help reduce anxiety. Many students who connect at revisit day stay in touch over the summer. When September comes around and school starts, they arrive on campus already familiar with the campus and some faculty, and having made a few friends. Even the most confident child is nervous about not knowing anyone at their new school. At the very least, attending a revisit day gives the student more confidence about entering the school as a new student.

Multiple Revisit Days

Most revisit days happen within a 5–10 day window in April. The concentrated time frame can make scheduling challenging for students who are considering several offers. Dates call fill quickly. Register early, particularly as an international family who has to make extensive travel plans to attend. When planning to participate in several revisit days, make sure to identify the factors that are important to you in your decision, and fully explore them during the revisit day. For example, if you are interested in robotics, try to spend some time in the robotics classroom at each school. Try to interact with the subject teacher and a student who has taken robotics classes, or one who may be on the robotics team. You are now the consumer, so make sure that the superstore has a prominent place for your brand.

Deposit and Contract

Upon confirmation of enrollment, parents will need to complete the enrollment documents and the required deposit before the date specificed by the school, usually on or around April 10th. Contracts will be sent at a later time to complete the enrollment process.

SPECIAL CONSIDERATIONS

Re-Application

Some students find that after having applied to boarding school, they have no offers. Gaining no acceptances can be disheartening. For some students, they will decide to stay at their current school and work towards preparing for US college or university. Staying at home can be the best choice for some students.

For other students, they are committed to attending a US boarding school. Perhaps it is the teaching pedagogy, a family tradition, or the desire to explore a certain academic subject or sport. For whatever reason, they plan to apply to boarding schools again.

In deciding to reapply, it is helpful to try to identify the reason the student did not get accepted. There are usually only a few reasons why this occurred:

1. *Not a good fit. Student is not academically or emotionally prepared to be successful at the boarding school(s) they applied to.*
2. *The student did not have a balanced school list.*
3. *The student did not develop or maximize their profile.*
4. *Financial aid was not available.*

Things to Consider before Re-Application

Academic Readiness

Students who have not demonstrated their ability to be successful at a boarding school academically will not be deemed a good fit for boarding school. The question becomes, is there enough time and motivation to improve this situation for a re-application, or is the student better served to remain at their current school, focus on the improvements, and prepare for a US university or college application process? Academic preparedness is also related to school selectivity. A student who is not deemed a good academic fit at one school may be seen as a good fit at another. Success could simply be a matter of re-examining the school list and being more realistic about fit.

Emotional Readiness

Students who have not demonstrated the emotional preparedness to face the challenges of boarding school are also not considered good candidates. This refers back to the Character Skills Snapshot and the 7 essential characteristics/ skills identified in the Snapshot. Students who are successful at boarding school demonstrate most of these skills to some degree before they enter boarding school. These skills prepare students to navigate the challenges of boarding school. For example, students who demonstrate self-control know when to prioritize schoolwork over having fun with friends. Students who understand the importance of teamwork and collaboration make great teammates, project partners, and dormmates. A student's ability to interact with others and adapt their behavior demonstrates social awareness. While initiative may be seen as motivation to push academically or on the sports field, a student who is feeling stress or homesick and takes the initiative to reach out for help is also demonstrating emotional preparedness.

Students who are admitted but not emotionally ready for boarding school can find themselves needing to return home. Returning home is the worst outcome for a student and a school because it affects students emotionally, and as mentioned earlier, the change of schools is reflected on the student's college/university application. Boarding schools are very careful to try to screen for emotional preparedness in the application process. In considering a re-application, does the student have time to try to resolve some of these issues? Are they capable of resolving them? Remember what was said in the beginning, not everyone is cut out for boarding school.

School List

Some families lack education about the competitive nature of the boarding school application process. They can be overconfident about the student's ability to get accepted and only apply to the most highly selective schools. Families look at rankings and don't consider what's the best fit for their child. Students must be aware of how they stack up to the competition and apply to schools of varying competitiveness. Students who are going to reapply need realistic expectations for their school list or history could repeat itself and they could again end up with no offers.

Once academic and emotional readiness is determined, a student's profile is evaluated to see where and how they can contribute. Building a profile that is competitive given the landscape takes time. Is the lack of offers a matter of the application simply not reflecting the student's X-Factors? Did the schools to which they applied recognize the X-Factors? Does the student simply not have enough to be a competitive applicant?

Families are not always aware of how well-developed the profiles are of other students who successfully apply. They are not able to see or accept the differentiation of their child. Other families recognize that their child may not be at the same level as their peers entering a given school but believe that "if" their child is just admitted to that school, their child will be motivated by their peers. Most top boarding schools are not looking for students who have the potential to be successful if admitted. Schools are looking for students who already demonstrate this!

Financial Aid

Boarding schools have limited financial aid budgets. Unfortunately, many qualified students are not able to get assistance or enough assistance to attend. For international students, financial aid can be even less available. A student who wishes to re-apply and will be applying for financial assistance, may want to consider applying to more schools and a wider range of schools.

Re-application is very sensitive and should be handled as such. Generally speaking, it is difficult to get admitted to a school that has already denied your application. For a re-application, students have from mid-March when they learn of their denial results until December when they are finishing their reapplication — essentially 9 months to try to turn things around. Students have been most successful reapplying to a new set of schools that either are more in line with their profile or can, at the very least, look at their application with a fresh set of eyes.

Take a deep breath, and enjoy your accomplishment. Being admitted to a US boarding school marks the beginning rather than an ending.

Students should make sure they continue to maintain their grades. Schools require students to send their end of the year transcript. If a student's grades significantly fall below the profile they were admitted from, the offer of admission can be revoked (Check the fine print of the contract!).

Spend time networking in your local community to find alumni and other students new and returning. Some schools will host admitted student receptions through their local parent chapters — attending them should be a priority. Embrace your school identity. Go online to the school store website and buy the school's t-shirt or sweatshirt that you were hesitant to purchase before you were admitted. Parents, don't hesitate to get a sweatshirt too. Parents are part of the community. The journey is just beginning for all!

Here is a timeline for the boarding school process. As was said in Step 1, "Start Early." It is recommended that students start at least 2 years before the year to intent to enroll.

Timeline	Activity	Reference
2 Years before Application	• Conduct a self-assessment to determine Academic Readiness -English Proficiency -Academic Performance -Critical Reading and Thinking Skills -Creativity and Problem Solving -Collaboration and Communication • Develop a plan to remediate any identified academic weaknesses or deficiencies • Enroll in test preparation and develop an overall Test Preparation Plan -SSAT -TOEFL • Build you brand -Create an Authentic Message -Pinpoint Your Mission -Decide How You Want to Be Seen -Research and Development of X-Factors -Think Like a Customer -Create Customer Loyalty • Promote your brand -Develop a Marketing Strategy -Develop a Social Media Presence -Have an Awesome Website -Be Involved in the Community • Build relationships -Schedule Preliminary Visits -Attend Local Receptions	Step 1.1 & 1.2

Timeline		Activity	Reference
Application Year	Jun.	• Evaluate your brand • Execute summer plan or program you may have to enhance or support your brand • Identify any standardized test weakness and develop a plan to improve • Identify a balanced preliminary school list • Identify when you will visit the schools • Request information from schools • Check email regularly	Step 2.1 & 2.2
	Jul.–Aug.	• Student to execute summer enhancement and standardized test plan • Finalize a preliminary list of (approximately 8–10) schools at which the student will interview • Request catalogues and review websites of the schools on your preliminary list • Complete online inquiries and pre-interview questionnaires • Contact schools to make appointment for official tour and interview • Complete Student Visit Schedule • Set up My Boarding School Research Tracker to keep track of school specific details • August 1: Register for SSAT • Check email regularly	Step 2.2 & 2.3

STEP 04

FOLLOW-UP ▸

Timeline		Activity	Reference
Application Year	Sept.	• Brainstorm and begin to draft application essays • Follow standardized test preparation plan • Begin Parent Statements • Check admissions travel schedule • Attend local receptions of schools you will apply to • Check email regularly	Step 2.2 and Step 3.1 & 3.2
	Oct.	• Take October SSAT • Continue to draft application essays • Follow standardized test preparation plan • Continue to draft Parent Statements • Identify any supplemental recommenders and verbally request recommendation (Personal Recommendation/Special Interest Recommendation) • Check email regularly	Step 3.2–3.4
	Nov.	• Take November SSAT if needed • November 1: Finalize School List • November 1: Schedule to meet with teachers in person to request required teacher recommendations (Math/English/Guidance Counselor) and final transcripts. • Complete Application Plan: SAO/GTP/School Specific). • Confirm any supplemental or personal recommendations (requests are based on Application Plan: SAO/GTP/School Specific) • Begin to finalize essays • Begin to finalize Parent Statements • Attend local receptions of schools that you intend to apply to • Check email regularly	Step 3.1–3.4

Timeline		Activity	Reference
Application Year	Dec.	• Take December SSAT if needed • Check on status of recommendations and transcripts; confirm that they were sent to schools • Finalize essays and review • Finalize Parent Statements • Take December SSAT if needed • Check email regularly	Step 3.2–3.3
	Jan.–Feb.	• Take January SSAT if needed • January 1: Submit standardized testing results • Submit all application materials before the deadline • Check school portals regularly for updates on status of application materials • Follow up with missing items as needed • Update your admission officer with any updates since submission • Check email regularly	Step 4.1
	Mar.	• March ~10th: Decisions will be released. Follow protocol recommended depending on the results. • Check email regularly	Step 4.2
	Apr.	• Attend revisit days of a few schools to which you were admitted. • Reply to schools with decision • Make a deposit to reserve a place at the school of your choice • Continue to do well academically • Check email regularly	Step 4.2

STEP

05 TRANSITION

5.1 – BEFORE YOU GO

5.2 - WHILE YOU'RE THERE

A transition signifies the passage from one stage to the next. When writing an essay, the transition helps the reader see connections between ideas, sentences, and paragraphs. In music, a transition links one music section to another. Moving from a student's current school to boarding school also represents a transition. The transition of changing schools is significant without the specific challenges of going to boarding school. Add the factors of living independently, sometimes thousands of miles from home, a new style of learning, adjusting to a roommate and new culture, figuring out how to make a successful transition to boarding school is crucial.

5.1 BEFORE YOU GO

Applying to boarding school is an intensely competitive process and for many families, getting in seems like it is the goal. Instead, boarding school should be viewed as a step in a student's educational journey. Like most meaningful experiences or adventures, there are several steps along the way, and successful journeys take preparation. Success at pretty much anything is a result of several factors. When defining success at boarding school, it, too, is determined by many different factors. First, students need to be willing and able to do the academic work required and to complete the work in a healthy way. Adaptability and resiliency are essential skills to help students deal with setbacks and failures. Students do best when they understand their strengths as well as their weaknesses and have a purpose and direction at their school. Acknowledging that asking for help is a sign of strength rather than a weakness is a strategy that benefits most students. Boarding school is rigorous; good time management skills and self-control help students to balance their commitments. Let's not overlook health and well-being because, without them, all of the above doesn't come together. Success comes when students make good choices and decisions, get enough sleep, and maintain a balanced diet.

Launching is difficult for both students and parents. There will be many emotions running through each party — nervousness, excitement, doubt, and worry. These are all normal.

If you have indeed started early, followed the recommendations related to Academic Readiness, Social and Emotional Readiness, and Extra-Curricular Readiness, you are likely ready to start this journey with your suitcase full of most everything you will need along the way. Don't fret. There will be bumps along any journey, but you have the resources to be able to overcome them because you have built the skills over time. Let's look at what you need to do to get ready.

ACADEMIC READINESS

Math Placement Test

The US mathematics curriculum is based on topics that are usually taught as a subject for a full year, such as Algebra I, Geometry, Algebra II, Pre-Calculus, and Calculus. Some international students, even those advanced in math, may

face challenges in math when entering a US boarding school. The difference is in the delivery of and exposure to the math curriculum rather than in aptitude or ability. Some international schools' math curriculums cover several topics within a specific grade level or class year, rather than a comprehensive study of a single subject within the discipline. It would not be unusual for an incoming 9th-grade international student to have learned some Algebra I, some Geometry, and a little Pre-Calculus. While some international students may have been exposed to more math topics than their US counterparts, they can have gaps in the complete knowledge of the overall subject matter.

Most boarding schools will require students to take a math assessment the summer before school starts. Students may want to consider brushing up on Algebra or Geometry so that they have a more in-depth understanding of the topics to prepare for the math assessment. Keep in mind that the school will also use the math recommendation to determine the student's math placement. Placing in advanced level classes will provide greater opportunities in future class selection, so keep your skills fresh over the summer.

Second Language Placement

There are usually several different levels of each second language offering at US boarding schools. If a student plans to continue a second language, schools will often provide them with a placement test in order to be placed at the appropriate level from the start. Students should also try to keep up their second language skills during the summer.

Class Selection

Most students entering grade 9 or 10 are going to have little choice in the core classes (English, Math, History, and Science) that they can take in the first year. They are likely going to be offered only a few electives to choose on their own. As students advance in grades and meet pre-requisites and graduation requirements of the school, they will be afforded more choice in class selection.

All-School Read

Over the summer, schools will usually assign a book that all faculty, staff, and students read. The subject of the book is usually something related to a current topic or tied to a school theme or initiative for the upcoming school year. All students reading the same book is intended to help students and teachers at the boarding school to strengthen their bond as a community. Discussing the book helps to build a shared understanding and experience to kick off the new school year. In addition to the All-School Read, grade specific summer readings often are assigned. Students must enter boarding school having read the books, done the assignments, and be prepared to discuss them. Often the discussion of these books is used as an activity during orientation. Make sure to start off on the right foot!

Fill in Academic Gaps

Some students find that spending the summer filling in academic gaps in subjects that are not necessarily their strengths gives them more confidence when starting at boarding school. Additionally, pre-learning subjects that students may be less familiar with may also give the student added confidence.

SOCIAL AND EMOTIONAL READINESS

Practice Independence

Many students can get derailed by the initial adjustment of being independent at boarding school. They have trouble keeping track of their assignments, getting up on time, and managing their self-care. This lack of organization can affect their academics and their overall adjustment and success at boarding school. For some students, these skills need to be learned and then practiced. Use the summer to try and build some of these skills. Set your alarm to wake up and fix your own breakfast. Learn how to wash and fold your laundry. Many boarding schools require kids to help with chores such as cleaning, meal preparation, and dishes. Learn how to use a vacuum, iron a shirt, how to prepare basic meals, and wash the family dishes. If you arrive at boarding school already ready to be independent, you will be able to focus more on academics, extra-curricular activities, and making friends.

Communication Skills

Boarding schools are based on the Harkness Method of teaching, where students discuss ideas in an encouraging, respectful, open-minded environment with only occasional or minimal teacher intervention. Students are expected to move the discussion forward by adding comments, asking, and responding to questions or supporting comments made by other students. This method teaches students to challenge the views of others by explaining their reasoning politely. Learning how to communicate is essential to boarding school success. Many students come from a teaching model that is less interactive than US boarding schools and find the transition very difficult. New students struggle with how and when to join a conversation. Some students may overly participate or fail to acknowledge the opinions of others. There is a learning curve here. If having not already done so, students may want to enroll in training or programs over the summer that encourage discussion-based learning.

Enjoy the journey, and don't forget that it is ok to ask for help!

EXTRA-CURRICULAR READINESS

All students need to participate in sports and extra-curricular activities at one level or another. If the school where a student is enrolling requires participation in sports, students should consider their options. If a student already has some experience in a sport that is offered by the boarding school, they may want to continue to participate in this sport when they attend. If so, keep up with the skills over the summer to help you prepare for tryouts when you arrive at school. For students who don't have a sport and want to try something new, they should do some research and find out what types of sports are offered at their school. Depending on how ambitious they are, they may want to familiarize themselves with that sport the summer before arriving on campus. Most schools offer different levels of sports teams that are suitable for all students, from the most competitive athletes to teams where students need no prior experience. Sports are a fantastic way to meet other students outside of the classroom. So even if a school does not require students to participate in sports, students might want to consider doing so, at least for the fall term.

Extra-curricular activities are another great way to meet other students, to explore your interests, or to find new ones. Students should have enrolled in a fit school, which means that some activities and interests match up with a student's interests. If you are interested in continuing something related to your X-Factor for example, find out when that activity meets and make sure to sign up. Students may find that they want to try something new, and that's fine too. The idea is to get involved and contribute to the school community. Once you adjust to the school, you may want to pursue activities or clubs that match your interests but do not exist. This is a great opportunity for you to think about starting your club.

The most successful international students at boarding school are the ones that broadly participate in both sports and extra-curricular activities which allow students to find a healthy balance between academic success and success in the boarding school community as a whole.

WHAT WILL I NEED?

Don't worry too much here. Schools will send a list of things that students need to start the year. Many families arrive a few days early to purchase the larger things that are on the list. Schools will also send students information about online companies which you can order from and will have the items waiting for you at school when you arrive. Most schools have some sort of big-box store/superstore (a real one not our imaginary one used for the purpose of this book) close enough by for you to purchase whatever you need or forgot. There is always the option to order items online and have them shipped to you at school, and there is usually a faculty-lead shopping trip to the local mall the first weekend on campus. No need to fret. Many important personal items can be purchased at the school store.

Packing List

Most dorm rooms will come furnished with the following: twin bed, clothing dresser, desk, chair, one garbage can and a closet for each student.

Boarding schools will send students a list of recommended items. Below are some basic guidelines for packing for international students heading to boarding school:

- *A family picture;*
- *Tech items: computer, phone, tablet and chargers;*
- *120-volt converter;*
- *Personal posters;*
- *Clothing: pending dress code requirements/sports clothing/casual clothing;*
- *First aid kit;*
- *Prescription medicine;*
- *Re-usable water bottle;*
- *Personal toiletries and shower shoes; and*
- *Personal athletic equipment.*

The following items you may want to consider purchasing when you arrive or order in advance:

- *Printer: check with your school for specific details (many schools are networked and this may not be necessary);*
- *Bedding: foam mattress topper, mattress pad, comforter, pillows, blanket and 2 sets of Twin XL sheets (most bedding is Twin XL, i.e., 39" x 80" x 4");*
- *Towels: 4 large bath and 4 hand;*
- *Bathrobe;*
- *Shower caddy;*
- *Window fan: many boarding schools don't have air-conditioning and September can be very hot in some areas of the US (fans with heaters for the winter are usually not allowed);*
- *Power strip;*
- *3M Command or Blue Tack for hanging up posters and pictures;*
- *Laundry basket;*
- *Hangers;*
- *Storage bins;*
- *Winter clothing: jacket, hat, mittens, boots, etc., depending on the climate;*
- *Hair dryer or any small electronics;*
- *Door mirror;*
- *School supplies; and*
- *Athletic equipment.*

The following items are usually not allowed and could be confiscated:

- *Halogen lamps;*
- *Televisions or large monitors (check school specific guidelines);*
- *Game systems (older students may be allowed);*
- *Wireless modems/routers;*
- *Air conditioners;*
- *Microwaves or mini fridges;*
- *Instapot;*
- *Heaters;*
- *Extension cord;*
- *Candles or incense;*
- *Weapons; and*
- *Plug in air freshener.*

Visa Application

You must have a student visa to study in the United States if you are not a citizen or green card holder. Some general guidelines are outlined here to give students and their parents an understanding of what is required, but visa consultancy can be complicated. Students with specific questions or situations should consult the resources provided below or an immigration attorney.

International, non-immigrant students cannot study to earn credit in the US on a B (visitor) visa. The course of study and the type of school determines whether international students need an F visa or an M visa. M visa is for a

vocational or other recognized non-academic program, other than a language training program. International, non-immigrant students intending to enter the US to attend boarding school will be required to obtain an F visa. Helpful US Government sites include:

EducationUSA https://educationusa.state.gov
Study in the States https://studyinthestates.dhs.gov
Student Visa Page of Travel.State.Gov https://travel.state.gov/content/travel/en/us-visas/study/student-visa.html

Student Acceptance at a SEVP Approved School

Any school that enrolls international, non-immigrant students with F visas must have approval and certification from the Student Exchange Visitor Program (SEVP). The SEVP is administered by the US government through the Department of Homeland Security (DHS) and Immigration and Customs Enforcement (ICE). SEVP approved schools go through a lengthy and intensive process when they first apply for certification. They need to show evidence of the content and quality of their programs, their overall services, their facilities, and their personnel. The certification process can take up a year, and during that time, the school will be visited by SEVP agents to observe every detail of the school. SEVP evaluates the knowledge of the school's personnel and the consistency of the school's records to determine if the school is fit to receive international students and gain approval for certification. And, even after the schools become certified, if the schools want to keep their certification over time, they need to operate under a very specific set of regulations all while being supervised by SEVP. SEVP approved schools need to assign one or more designated school officials (DSO) and one principal designated school official (PDSO), who must undergo specific training to meet SEVP's qualifications.

Boarding schools that accept non-immigrant international students have applied and have met the requirements to be a SEVP approved school. After the SEVP-approved school processes a student's enrollment including tuition fees, students will be registered for the Student and Exchange Visitor Information System (SEVIS) and must pay the SEVIS I-901 fee. The SEVP-approved school will issue a Form I-20. After the student receives the Form I-20 and registers in SEVIS, the student may apply at a US Embassy or Consulate for a student (F) visa. The student must present the Form I-20 to the consular officer when attending the visa interview.

Plan ahead because it can take time to get appointments at the US Embassy or Consulate in the country where you live. For new students, the F visa can be issued up to 120 days in advance of the start date for a course of study. However, you are not allowed to enter the US on your student visa more than 30 days before the start date of your school.

5.2 WHILE YOU'RE THERE

A good understanding of how a boarding school operates and how to best utilize the wide array of available resources is like a road map, helping students to navigate their journey. Boarding schools bring together students from many different backgrounds and experiences to share in a common community. Boarding school is a time to learn to live independently of your parents but in partnership with other adults in the community.

Each student is assigned a faculty advisor when they arrive on campus. Students who embrace the advising system build lasting relationships that can sustain and support them throughout boarding school and often long after. Understanding how to meet the curriculum requirements and seeking advice on how to choose higher-level courses that are meaningful and challenging prepares students for successful college placement. Living with others in the dorms teaches life skills, such as tolerance and cooperation. Throughout this journey, students will learn about their strengths and how to ask for help in areas of challenge. They become healthy and active young adults participating in sports and exploring their passions and interests.

ADVISING

All boarding schools have an advisor system. As mentioned above, once enrolled, students will be assigned an advisor. Some schools ask advisors to reach out to students before arrival, and other schools have advisors introduce themselves on arrival day. Schools try to match initial advisors to students based on some regularly planned interactions. A student's advisor is usually a member of the school that the student regularly interacts with, such as a teacher, coach or dorm parent. Dorm parents are adults who have a separate apartment within the dorm and essentially look after the students in their dorm or hall.

The role of the advisor is to guide the student and their parents through the boarding school process. The Advisor-Student relationship can be one of the most meaningful at boarding school. Advisors serve as mentors who help students navigate through the inevitable difficulties that will arise and celebrate in their successes. Advisors are familiar with the requirements and curriculum and can be very helpful in assisting students to map out their academic plans. Advisors are always the first point of contact for the students and their parents.

COURSE SELECTION

Every school has its curriculum requirements that form the core of the graduation requirements. Students should familiarize themselves with the curriculum and the requirements, such as how many credits of different subjects a student will need in order to graduate. Knowing what academic requirements are needed helps avoid the situation of a second-term senior who discovers they need another art credit to graduate! Students should "look ahead" to identify any prerequisites required to take higher-level courses or electives in grades 10–12. Students should devise an academic plan with their advisors for each year at the school. This document can serve as a reference, and students and advisors can adjust as student's goals change.

RESIDENTIAL LIFE

It would be unusual for boarding school to be a student's first experience being away from home, particularly for an international student. As part of the admission process, most boarding schools are looking for some experiences where students have lived away from their parents, whether it be a summer program for a few weeks or an extended school trip. Many students initially feel that boarding school will be a little like summer camp. That is true for a few weeks, but as the routines set in, boarding school becomes less of a festive atmosphere most of the time. Living away from home as a young person can be difficult. Students can get homesick or have a conflict with their roommate(s), or both. They may be challenged by simple tasks like keeping their rooms clean for room check or managing their technology time.

The key to adjusting to residential life goes back to the Characters Skills (Step 3.2). In particular, the skills of open-mindedness, self-control, and social awareness are important. Boarding schools value and promote diversity. Students who have an open mind about other students' backgrounds and experiences are more likely to be successful in the residential setting. Students may find that they are living with a person who is very different than themselves. Keeping an open mind will help to promote acceptance of those differences rather than viewing them in a divisive way. Good living partners

and roommates practice good self-control. They try their best not to negatively impact their roommates and are cooperative living partners. Good roommates try to be considerate of how they live within the confined space of their room and respect their roommate(s). Students who are successful in a residential setting are aware of the social cues that exist when we are overstepping our boundaries or getting on the nerves of others.

Despite best intentions, there are times when rooming situations result in an unrepairable conflict or unhealthy situation for all roommates. These are the times where students need to learn the skill of self-advocacy and find an adult to help them navigate the conflict, like their advisor or dorm parent.

SELF-ADVOCACY

Self-advocacy is an important life skill, and boarding school is a fantastic place to begin developing it. Self-advocacy is a term used to describe the act of representing yourself. Self-advocacy enables students to know their strengths and weaknesses, understand what they require to succeed, and how to communicate this to others. In a nutshell, self-advocacy is the ability to speak up about what you need in a way that promotes the delivery of what you need. Being a good self-advocate is a skill that requires nurturing and leads to success in boarding school and in life. The adults at boarding school will not always know that you are having difficulty. Whether you do not understand a reading assignment in your philosophy class or feel sad and lonely because you are homesick, let an adult know. A solution can usually be found to most challenges. At boarding school self-advocacy translates, quite simply, into asking for help if you need it!

SPORTS AND EXTRA-CURRICULAR ACTIVITIES

Becoming involved in the community through participation in clubs and activities is one of the hallmarks of boarding school. Students who embrace involvement are more likely to be successful. Make sure to meet any athletic or extra-curricular requirements. Identify sports, extra-curricular activities and clubs that interest you. For some students, consistent involvement in a few key clubs or activities often opens the door to leadership roles at the school. Keep in mind that you are on an educational journey, and your boarding school journey will lead you to your university journey. Keep working on your X-Factors or develop new ones. Some students re-brand themselves entirely in boarding school, while others work to refine and strengthen their original brand.

SUCCESSFUL INTEGRATION INTO US BOARDING SCHOOL CULTURE

Every boarding school has a mission statement that directs their culture. Students will be reminded of the mission regularly as it informs many of the school's decisions and policies. One school's mission statement may speak to diversity and inclusivity while another's sets to inspire the best in each to seek the best in all. Students should know and understand their school's mission statement and look for opportunities to demonstrate that mission at their school. Boarding school culture is about the greater good, teamwork, and cooperation.

On a practical level, successful integration into boarding school culture means getting involved in the community. Students can choose to isolate themselves in their rooms. Or they can join clubs and participate in activities that enrich their own experiences and that of the entire boarding school.

International students face an even more significant challenge to also integrate into American culture while maintaining their own cultural identities. Getting involved in school activities and sports will help international students to meet more American students. Most boarding schools also have cultural affinity groups that are available to help students spend time among students with whom they may share common cultures and experiences. Students who are most content at boarding school embrace students who are different than themselves, and find time to maintain their cultural identity by engaging with students of similar backgrounds.

PARENT INVOLVEMENT

Boarding schools welcome parent involvement and see parents as partners in preparing the student for success. Parents are welcome to attend athletic games and to visit campus. For parents of most domestic students, this is an easier task than for most international families. Schools provide opportunities for all parents to visit campus, like parents' weekends. Some schools stream performances and athletic events live or record them so parents can view from their home countries. The internet affords today's parents many opportunities to be involved. Parents can send cupcakes to the dorm for the child's birthday or sponsor refreshments for the team at a sporting event — all with the click of your mouse or a call to the advisor.

For parents, the student's advisor is their point of contact at a school. Most schools have a formal protocol by which advisors must interact with parents. This is usually around the distribution of grades and comments. Advisors are readily available to speak/meet with parents who may have some concerns or are simply looking for reassurance. Developing a good relationship with a student's advisor is always beneficial for parents and, ultimately, their child.

There is always a balance in this partnership. Parents who are too involved can unknowingly impact their child's experience. Parents need to trust that the school has the student's best interest in mind. Both parents and schools share a common goal — helping the child develop into a responsible global citizen.

School Transfer

A transfer from one boarding school to another is not standard, but it does happen for good reasons. School transfers are most often quite complicated and should be avoided if possible. School transfers will be evident during the college application process and can raise a red flag.

What are some of the reason why a student would transfer to another boarding school?

1. Academic Fit

Sometimes students' academic strengths can be misinterpreted in the admission process, and the students find themselves in a situation where they are not able to keep up despite their best efforts. Students should take all necessary steps to make sure that their next school is an appropriate academic fit and explore resources available to provide extra support while they transition.

2. Discipline Issues

Schools are full of young adults who are exploring and testing limits. Each year, boarding schools will have a few students who are not fit for the school's culture based on their behavior and actions. These students are either dismissed immediately or are not asked to return for the following year.

Boarding schools have very strict rules, which help promote a safe and equitable community. It is worth exploring each boarding school's disciplinary policy before identifying it as a school on a student's list. Some schools may have a "one-strike" policy for certain infractions, such as academic dishonesty, drugs and alcohol use, or violence. This means that a single infraction can result in dismissal. Others may have a policy where students are placed on probation and given restrictions after a first offense. A second offense often results in dismissal.

If dismissed from school or not invited back the following year, students are left with a decision about where they will complete their education. With a tarnished record this can be very challenging. For those students who wish to continue to attend a boarding school, this becomes a 2-step process.

1. *Did I resolve the issue that I was disciplined for?*
2. *How do I demonstrate this to give my new school confidence that I can obey the rules at their school?*

Some schools may be willing to give a student a second chance, admitting him/her to their school if the student demonstrates remorse and having learned from or resolved whatever disciplinary infraction occurred. This type of application can be very sensitive and should be handled as such.

CASE STUDIES

Often, the best way to illustrate a point is to give an example; a demonstration of sorts. Throughout this book, the student, as the brand was discussed. The importance of identifying, developing, and demonstrating brand attributes to boarding schools was examined in depth. Now it is time to see how **Start Early**, **Plan Well**, **Execute**, **Follow-Up**, and **Transition** can prepare students to be admitted to some of the most competitive boarding schools in the United States.

A demonstration focuses on the rewards or benefits of brand-specific features to get potential buyers excited about the product. A product demonstration provides visual support to grasp the product's value and potential fully. Instilling a sense of ownership of the product is one of the benefits of product demonstrations. Product demonstrations also support concrete evidence of their claims. Lastly, demonstrating a product helps to address product-related concerns; a demo can ease those worries.

In this section, Jennifer Yu Cheng, Co-Founder of ARCH Education will provide a few Case Studies as product demonstrations.

Like the students in our Case Studies, others can also increase their chances of finding their brand on a boarding school shelf by following 5 Key Steps:

Step 1 START EARLY

Step 2 PLAN WELL

Step 3 EXECUTE

Step 4 FOLLOW-UP

Step 5 TRANSITION

CASE STUDY: "MICHAEL," THE MAKEOVER

Michael arrived at ARCH an above-average seventh-grade student, two years after relocating to Hong Kong from the US. He was a student who had gotten stuck "in transition," adjusting from the expectations of an American suburban school system to those of an international IB school system. He had become overly comfortable coasting through school. He essentially had no authentic message or inspiration for his brand.

Our in-person impression of Michael was that he was immediately much more interesting than his file. It became apparent that Michael was extremely driven and motivated, deep inside, to do well — he just needed the right feedback and guidance to leap forward. He needed to pinpoint his mission; set a goal — and his goal became to go to boarding school. Once he set this goal in his sights, he began to seriously fulfil his potential in school and extra-curricular activities, specifically in sports and music.

While Michael would not be the best candidate to demonstrate the importance of Step 1: Start Early, we identified important readiness indicators just below the surface. Michael had only a small window — nine months — to turn things around before applying to boarding school. He needed to commit fully to the process to demonstrate his full potential. Michael used this incentive — of getting accepted to boarding school — to dramatically improve his grades. On a 7-point grading scale, he moved from an average of 5.4 to an average of 6.4 in just 9 months. He was so excited about the prospect of boarding school that it was easy to encourage him and set new expectations for him.

Michael is a classic case of complacency transformed. With the right goal, motivation, encouragement and mindset, Michael was able to enrich his potential dramatically. His SSAT score landed him in the 99th percentile. Michael also joined the rowing club at school, auditioned for and was accepted into a citywide youth orchestra, and capitalized on a newfound love of law and humanities to engage in different debating and public speaking opportunities, including attending a US based summer program the summer before his application. He went through the process of research and development and identified his X-Factors.

Through goal setting, Michael transformed across the board within a short nine-month period and reached his target: being admitted to a top 10 US boarding School!

CASE STUDY:
"CHRIS," THE PLANNER

Chris arrived at ARCH as a student in our enrichment program in 6th grade. He was a strong writer and sought opportunities outside of school to advance his writing. He enjoyed the dynamic, discussion-based learning classroom approach at ARCH which mirrors US boarding schools. Through the ARCH experience, he and his family quickly became interested in exploring education options in the US.

Chris would be considered an "early planning" case. What truly sets Chris's case apart is how early he and his family lay the groundwork for his education in the US under the guidance of ARCH's admissions consultants. Chris soon found himself exploring new extra-curricular interests and attending summer programs in the arts, writing specifically. Chris was young for his grade and thus his consultant recommended a two-step transition — from junior boarding school to boarding prep school — to allow him to gain confidence, mature, and explore leadership opportunities. Over time, his Emotional Readiness developed.

ARCH helped Chris tap into his full potential as a writer, artist, and as a leader (X-Factors). Through our small classes, he was able to fully explore his writing skills and talents. He explored summer programs in fashion design and arts prior to applying to junior boarding school. In 8th grade, he was accepted to junior boarding school, where he grew up quickly. By the time he graduated junior boarding school in Grade 9, Chris was already on a Varsity sports team, chief editor of his school newspaper, and was even elected as Student Council President. Had Chris stayed in Hong Kong, he simply would not have had as many opportunities to fulfil his full potential.

Chris had always been academically strong, but the extra year at junior boarding school helped cement his confidence and his sense of self. Furthermore, Chris ended up choosing a school not according to ranking but according to what was right for him specifically. From his junior boarding school experience, he knew that school fit was an important factor and tied to student success.

At ARCH, due to early planning, we were able to encourage him to explore the full breadth of his activities and interests and to take on leadership roles — all before applying to boarding school. When he started boarding school in 10th grade, Chris hit the ground running and started a school publication, played in the school orchestra, and became a leader on the student affairs committees, to name a few areas of extra-curricular involvement; future X-Factors for his university application.

By the time Chris was a junior, he had already identified his strengths and weaknesses through and through. By senior year, he was pursuing the right opportunities for himself — to tap into his true passions — and was accepted to an Ivy League university.

CASE STUDY:
"NATALIE," THE EXPLORER

Natalie arrived at ARCH a soft-spoken, slightly reserved but academically strong 8th grader. She wanted to explore boarding school, as she felt limited by the options available at her local school. She exhibited some interest in extra-curricular activities like playing the flute, but was unclear about her passions and future academic direction. She indicated that she liked reading, writing and filmmaking — but hadn't yet explored any of these in depth.

After getting to know Natalie a bit, we thought that going to boarding school would benefit her by providing a much bigger platform on which to explore her interests in and out of the classroom. Especially with her interest in humanities, boarding schools offered a much wider range of elective courses. We also knew that to be accepted to a top boarding school, she would have to overcome certain limitations such as speaking confidence and the ability to contribute to class discussions. Natalie spoke very softly, which gave an impression of low self-confidence; but as we got to know Natalie better, we came to know her as a resilient and determined student eager to self-improve. Natalie was determined to improve her speaking voice and confidence. We set out to work with her on improving her speaking skills — her projection and articulation. After working hard to improve her speaking skills, her self-confidence shot up. Altogether, the entire preparation process for her was truly transformative. The shy Natalie who entered our offices was not the beaming Natalie accepted to a top 10 US boarding school!

Part of the key to Natalie's application success was also her willingness — and that of her parents — to apply to a US boarding school as a repeat 9th grader. This single step afforded her an additional year to transition from local school curriculum in Asia to US curriculum standards, and to explore the full gamut of extra-curricular activities. And explore she did!

Once Natalie got into boarding school, she hit the ground running. She founded her own poetry club, and spent most of her extra-curricular time writing for school publications. She also took up step dancing! The magic of performing, of stepping far outside her comfort zone, awakened a newfound confidence in her. She felt so empowered through dance that she started a charity at home to uplift under-resourced children through step dancing.

Ultimately, Natalie became a very involved member of her school and her community. She won a series of writing prizes and was eventually accepted to one of the HYPS* universities! Had she stayed in her local school, she would not have had the resources she needed for her personal growth. She was truly transformed by the entire experience from start to finish.

CASE STUDY:
"ADRIAN," THE DO OVER

Adrian is an interesting case, because he came to ARCH as a 9th grader after being rejected and/or waitlisted by all the boarding schools he had applied to on his own the year before we met him. He approached ARCH to see if he had any chance of getting off a waitlist or if he should reapply to the boarding schools again in the coming fall. Our first impression was that Adrian was really mature, diligent and talented — but after reviewing his previous application, we felt that his application did not reflect any of this. Rather, it echoed only what he thought the schools wanted to hear. In our view, Adrian did not accurately align his X-factors with what the boarding schools were looking for. In other words, he didn't know his customer. Adrian was motivated to reapply, so we decided to redo all of his boarding school applications to fully portray and reflect what he had to offer. During this time, we had the opportunity to get to know Adrian well and work closely with him

*HYPS = Harvard, Yale, Princeton, Stanford

to tap into his true talents and interests, which included history, sports and music. With our deep knowledge of extra-curricular programs, we were able to suggest the right opportunities that would help him further develop and excel his talents, allowing him to shine.

Adrian attended an international school that didn't offer many competitive sports. In preparation for his boarding school transition, we encouraged him to expand his sports involvement and introduced him to new athletic activities — crew and cross country. We also discovered that he is a natural singer! A baritone with a rich, full voice — but he had never had any formal training. We took it upon us to introduce him to a professional opera instructor, and he ended up submitting his vocal recording along with his boarding school applications. Adrian was also an enthusiastic history buff and led his school's history team to the championships — but he hadn't fully conveyed this fact in his first round of applications.

Adrian was a classic case of hidden potential and a misguided application. Without counseling, we don't believe he would have leapt to the next level. We helped him explore the entire realm of possibility of studying in the US, which allowed him to discover amazing schools that had so much to offer him specifically. Adrian landed himself several offers, one at the top 10 school that previously waitlisted him!

EDUCATIONAL CONSULTANTS

There is nothing more essential to ensure a successful future for your child than a good education. The investment in resources is considerable, and the best way to help ensure success when making any investment is to hire a professional. In this case, **Educational Consultants** are professionals who advise students and their families considering a boarding school education and can be critically important to its success. Their first priority is to have the wholistic goal of what is best for the student. Hiring an Educational Consultant can help in the decision-making and planning process. Beware, not all Educational Consultants are the same. Founded in 1976, the Independent Educational Consultant's Association (IECA) has nearly 2000 members with representatives in every state in the United States and in 29 other countries. To be a member of IECA, Educational Consultants must meet standards for education and training. They are skilled professionals who visit schools annually to ensure their knowledge base is up to date. Further, IECA members must adhere to a code of ethics for the benefit of their clients and the schools to which their clients apply.

Educational Consultants are experts at understanding what is a very complicated and time-consuming process — applying to boarding school. Some having worked with boarding schools, they are adept at understanding the types of students boarding schools are looking to admit. Parents should view Educational Consultants as objective partners in helping prepare students not only to be admitted to boarding school but to be successful once the students are admitted. Hiring an Educational Consultant can help ensure a successful outcome to the boarding school process and beyond.

If a family is considering hiring an Educational Consultant, they should strongly consider only hiring one that is a member of IECA who has years of experience and a proven track record. A list of Educational Consultants who are members of IECA can be found on the IECA website: **iecaonline.com.**

USING AN EDUCATIONAL CONSULTANT

Planning and Advising

Educational Consultants help families navigate the complicated boarding school process from start to finish. As you learned earlier, the importance of starting early in the process of preparing for boarding school cannot be underestimated — the entire Step 1 was dedicated to this message! Engaging an Educational Consultant early in the process is consistent with that message. Hiring a consultant early in the process can help families by giving them an objective and clear evaluation of their child's strengths and weaknesses as they relate to preparedness for the boarding school application and future boarding school success. More importantly, based on this objective assessment, Educational Consultants work with families to maximize their child's strengths and identify resources to assist in overcoming the child's challenges. Educational Consultants also work to help identify the student's interests and passions and make recommendations on how the student can explore them further. Working with students over time, allows Educational Consultants to get to know their clients, and can help families evaluate students' character skills and make further recommendations on experiences to help further develop or enhance these skills. Not surprisingly, many parents find their children more receptive to the recommendations/directions of an Educational

Consultant, than if they themselves make the suggestions. Working with a consultant can be welcomed support for parents challenged by being the only adults committed to motivating and communicating with their child throughout the boarding school process. Finally, the result of all this work will allow the Educational Consultant to play a critical role in helping families develop a comprehensive plan to navigate the boarding school process.

School List

Educational Consultants who specialize in boarding school counseling should be well informed of the school options. They visit schools regularly, understand the boarding school cultures, and keep up to date on programs and offerings of the different boarding schools. Educational Consultants get to know the students that they are working with. In partnership with the student and family, they can provide an informed and balanced school list that best fits the student and family expectations.

Application Management

In the business plan for applying to boarding school, the Educational Consultant is like a "project manager." They are experts in the various application methods used by different schools and the nuances of helping identify the student's strengths in preparing the specific school applications, whether it be an SAO, Gateway, or School Specific Application. They understand the timelines and necessary components for an application, which helps their students shine.

Reliable Advice

Many families base significant decisions on the advice of friends. Would you base an important medical decision on the experiences of a friend who has similar symptoms, or would you go to a doctor to seek expert advice? While not life-threatening, education decisions are nonetheless important for your child's future. Therefore, it is highly recommended that you seek a professional to advise you on education decisions like you would a doctor on medical decisions. As an Educational Consultant, I have received many emails from parents that started with, "My friend whose son applied to boarding school told me I should....." Often, the friend's information is inaccurate, or the friend's particular situation does not fit all students, particularly your child! Parents should have an objective partner to inform them and assist throughout the boarding school process, rather than a friend whose only base of knowledge is their own personal experience.

Parents also rely on social media where the sources and intentions are never as transparent as they should be and may not be intended to help. As parents, how often have you reminded your children to check the sources of the information they read online? Are the sources reliable? Objective? Parents should heed this advice when it comes to making decisions about the information they read online related to the boarding school admission process. Educational Consultants are professionals trained in the admissions process and are hired by families to provide accurate and objective information.

COMMON QUESTIONS ABOUT EDUCATIONAL CONSULTANTS

What is the difference between an Educational Consultant and an Agent?

There is a big difference. An Educational Consultant, inducted into IECA, is a professional who has met strict educational and professional requirements. They are also ethically bound not to accept compensation from the schools at which they place students. The nuance of this last piece of information can be further explained in that IECA Members are engaged by the family, not the school, and inherently have the student's best interest as their primary focus.

Agents, on the other hand, have no professional organization that scrutinizes their experiences and monitors educational training. They are not held to any professional standards. Further, some agents are paid by both the schools and the families, creating a potentially serious conflict of interest. There are stories of some agents promising and even guaranteeing unrealistic admission results. Educational Consultants are working towards helping the student not only get admitted to the best fit school but to succeed throughout their boarding school experience. Admission to boarding school can never be guaranteed. If someone is charging a fee and insuring acceptance, you may want to consider hiring someone else.

Will it harm my child's chance of being admitted if the school finds out he/she worked with an Educational Consultant?

On the contrary, most schools welcome the involvement of an Educational Consultant. In fact, most reputable and experienced Educational Consultants will have a well-established rapport with the boarding schools. They have worked with them for years placing students and have established a trusting relationship with many of the schools. Many admission officers find comfort knowing that a student is working with a reliable Educational Consultant because they know that the consultant's agenda is about preparation and fit, rather than simply getting a student admitted. Many boarding schools are welcoming and receptive to input from Educational Consultants working with students who apply to their school. That said, in the end, families have a choice of informing a boarding school that they are working with an Educational Consultant or giving permission to the consultant to speak to the schools about their child.

Many families find having an Educational Consultant guiding them in this process to be critically important. Ultimately your child will be admitted to a boarding school on their own merit and the best chance of gaining admission is to prepare for the experience.

To get best value from this book, simply scan the QR code to download ***MY BOARDING SCHOOL PLAN WORKBOOK.***

CONCLUSIONS AND ACKNOWLEDGEMENTS

This book is the culmination of many years of experience working with international students to apply to the most prestigious boarding schools in the United States, as both an Independent Educational Consultant and as the head of international admissions for a top US boarding school. However, my most cherished boarding school experience is that of a parent of 4 boarding school educated children: Caitlin, twins: Claire and Elizabeth (Libby), and Drew. Caitlin and Claire attended the same boarding school. Elizabeth (Libby) and Drew each attended different boarding schools; 4 children, 3 different boarding schools! Drew, the youngest and most adventurous of the 4, has shown me that there is no full stop at the end of a sentence when pursuing your interests at boarding school. Caitlin, Claire, and Libby are thriving in their careers, having earned advanced degrees in their professions. Special thanks to Libby for providing all of the book's artwork while working full time as a Kindergarten teacher. Also, thanks to my husband, Tom, the product of a boarding school education, for his thoughtful edits and for showing me early on in our relationship the importance of the bonds with his boarding school and classmates, many of them he remains in contact with to this day. Participating in each of my children's boarding school education reaffirmed to me as a parent and as a professional the importance of fit and how allowing young adults (my own children and students that I work with) to make their own choices increase the likelihood of a successful experience.

Special thanks to Jennifer Yu Cheng, Co-Founder of ARCH Education in Hong Kong, who has been my partner in this journey since 2010. Jennifer herself was educated in a top US boarding school and Ivy League US university. In addition to contributing to the **Case Studies** Special Section, Jennifer has been a friend and mentor, allowing me to help her build a successful educational services business. She approaches every decision from the vantage point of what is best for the student and has the unique ability to fully engage our students to explore their passions and expect the best of themselves. Together with our team at ARCH, Jennifer and I have helped hundreds of students to realize their dream of attending US boarding school. Equally fulfilling for our students has been the realization that the opportunity of attending a US boarding school is more than just the academic education.

My aim in writing this book, is to provide general education and extra-curricular guidelines for both students and parents on how to approach the boarding school process. It is worth restating that the key is differentiation, so use this book as a resource and as a guide to help you explore yourself. Find areas of interest that you are passionate about and pursue them, not just for your boarding school application but for you, as well. Applying to boarding school should be a self-reflective process. Whether you end up applying to boarding school or not, reading this book and completing the worksheets can help you develop more as a student and an individual. **Start Early!**

The process of preparing for US university is very similar to that of US boarding school. Students who read this book and decide that US boarding school is not for them or are not yet prepared but wish to attend college or university in the US can still use most of the principles covered in this book to help them prepare for their college application. US Universities will look for similar qualities and differentiation in the admission process.

Our ARCH Education University Team guides students in a similar way as has been identified throughout this book. We work with students not only to pursue US university admissions but also UK and Hong Kong Universities. As I hope you have gathered from the pages of this book, there is a great deal of thought, dedication, and planning involved in being admitted and transitioning to boarding school. A framework is provided. What is most important is how a student approaches each of these steps. Seeking professional advice from an Educational Consultant on how to best manage and plan for the US boarding school process can help ensure success. Ultimately students gain admission on their merit.

Attending boarding school is transformative for many students, including my children, and the friendships and relationships fostered often last a lifetime. As we continue to work with many of our boarding school students through ARCH Education, helping them to apply to the most prestigious universities in the US and UK, we are reminded of the important role and awesome responsibility we are given to help families plan and make the best educational choices for their children. Thank you to the many families in Hong Kong and worldwide who have trusted ARCH Education with this important responsibility. For more information about ARCH Education, our transformative enrichment and academic programs, and US and UK boarding school and university consultation, please visit our website at **http://www.arch-education.com.**